Lecture Notes in Artificial Intelligence 11406

Subseries of Lecture Notes in Computer Science

More information about this series at http://www.springer.com/series/1244

Martin Atzmueller · Alvin Chin ·
Florian Lemmerich · Christoph Trattner (Eds.)

Behavioral Analytics in Social and Ubiquitous Environments

6th International Workshop on Mining Ubiquitous and Social Environments, MUSE 2015
Porto, Portugal, September 7, 2015
6th International Workshop on Modeling Social Media, MSM 2015
Florence, Italy, May 19, 2015
7th International Workshop on Modeling Social Media, MSM 2016
Montreal, QC, Canada, April 12, 2016
Revised Selected Papers

 Springer

Editors
Martin Atzmueller
Tilburg University
Tilburg, The Netherlands

Alvin Chin
BMW Technology Group
Chicago, IL, USA

Florian Lemmerich
RWTH Aachen University
Aachen, Germany

Christoph Trattner
University of Bergen
Bergen, Norway

ISSN 0302-9743   ISSN 1611-3349   (electronic)
Lecture Notes in Artificial Intelligence
ISBN 978-3-030-33906-7   ISBN 978-3-030-34407-8   (eBook)
https://doi.org/10.1007/978-3-030-34407-8

LNCS Sublibrary: SL7 – Artificial Intelligence

This Springer imprint is published by the registered company Springer Nature Switzerland AG
The registered company address is: Gewerbestrasse 11, 6330 Cham, Switzerland

Preface

The emergence of ubiquitous computing has started to create new environments consisting of small, heterogeneous, and distributed devices that foster the social interaction of users in several dimensions. Similarly, the upcoming social web also integrates the user interactions in social networking environments. However, the characteristics of ubiquitous and social mining in general are quite different from the current mainstream data mining and machine learning. Unlike in traditional data mining scenarios, data does not emerge from a small number of (heterogeneous) data sources, but potentially from hundreds to millions of different sources. Often there is only minimal coordination and thus these sources can overlap or diverge in many possible ways. Steps into this new and exciting application area are the analysis of this new data, the adaptation of well-known data mining and machine learning algorithms, and finally, the development of new algorithms. Behavioral analytics is an important topic within ubiquitous and social environments, e.g., concerning web applications as well as extensions in mobile and ubiquitous applications, for understanding user behavior. Some important issues in this area include personalization, recommendation, tie strength or link prediction, community discovery, profiling by extracting and understanding user and group behavior, predicting user behavior and user modeling, and prediction from social media.

This book sets out to explore this area by presenting a number of current approaches and studies addressing selected aspects of this problem space. The individual contributions of this book focus on problems related to behavioral analytics in social and ubiquitous contexts. We present work that tackles issues such as natural language processing in social media, collective intelligence, analysis of social and mobility patterns, and anomaly detection in social and ubiquitous data.

This book is based on submissions which are revised and significantly extended versions of papers submitted to three related workshops: The 6th International Workshop on Mining Ubiquitous and Social Environments (MUSE 2015) which was held on September 7, 2015, in conjunction with the European Conference on Machine Learning and Principles and Practice of Knowledge Discovery in Databases (ECML-PKDD 2015) in Porto, Portugal, the 6th International Workshop on Modeling Social Media (MSM 2015) that was held on May 19, 2015, in conjunction with ACM WWW in Florence, Italy, and the 7th International Workshop on Modeling Social Media (MSM 2016) which was held on April 12, 2016 in conjunction with the ACM WWW in Montreal, Canada. With respect to these three complementing workshop themes, the papers contained in this volume bridge the gap between the social and ubiquitous worlds.

The first paper is "Link Classification and Tie Strength Ranking in Online Social Networks with Exogenous Interaction Networks" by Mohammed Abufouda and Katharina Anna Zweig. The authors address the link assessment problem in social networks that suffer from noisy links by using machine learning classifiers for assessing and ranking the links in the social network of interest using the data from the exogenous interaction networks.

In the second paper, "Stratification-Oriented Analysis of Community Structure in Networks of Face-to-Face Proximity" by Stefan Bloemheuvel, Martin Atzmueller, and Marie Postma, the authors discuss performing automatic detection of face-to-face proximity during two student meet-ups for the purposes of community detection using wearable sensors and analyzing the proximity interactions between the students to indicate social relationships.

In the third paper, "Analyzing Big Data Streams with Apache SAMOA" by Nicolas Kourtellis, Gianmarco De Francisci Morales, and Albert Bifet, the authors deal with analyzing big data streams using a new open-source platform called Apache SAMOA (Scalable Advanced Massive Online Analysis).

The fourth paper entitled "Multimodal Behavioral Mobility Pattern Mining and Analysis using Topic Modeling on GPS Data" by Sebastiaan Merino and Martin Atzmueller deals with identifying risky driving behavior to increase traffic safety by using topic modeling and mobility pattern mining on GPS data.

In the fifth paper, "Sequential Monte Carlo Inference based on Activities for Overlapping Community Models" by Shohei Sakamoto and Koji Eguchi, the authors present their work on an incremental Gibbs sampler based on node activities that focuses only on observations within a fixed term length for online sequential estimation of the Mixed Member Stochastic Blockmodel.

The sixth paper, "Results of a Survey about the Perceived Task Similarities in Micro Task Crowdsourcing Systems" by Steffen Schnitzer, Svenja Neitzel, Sebastian Schmidt, and Christoph Rensing, provides an empirical study about how workers perceive task similarities in micro-task crowdsourcing systems based on cultural background, worker characteristics, and task characteristics.

Finally, the seventh paper is "Provenance of Explicit and Implicit Interactions on Social Media with W3C PROV-DM" by Io Taxidou, Tom De Nikes, and Peter M. Fischer. The authors discuss how with the enormous amount of social media data that is generated quickly, it is difficult to determine the relevance and trustworthiness of the information in that data, therefore they created a model to tackle this based on the W3C PROV Data Model.

We hope that this book (i) catches the attention of an audience interested in recent problems and advancements in the fields of big data analytics, social media, and ubiquitous data and (ii) helps to spark a conversation on new problems related to the engineering, modeling, mining, and analysis in the field of ubiquitous social media and systems integrating these. We want to thank the workshop and post-proceeding reviewers for their careful help in selecting and the authors for improving the submissions. We also thank all the authors for their contributions and the presenters for their interesting talks and the lively discussions at the three workshops.

July 2019

Martin Atzmueller
Alvin Chin
Florian Lemmerich
Christoph Trattner

Organization

Program Committee

Luca Aiello	Bell Labs Cambridge, UK
Christian Bauckhage	Fraunhofer, Germany
Martin Becker	Universität Würzburg, Germany
Alejandro Bellogin	Universidad Autónoma de Madrid, Spain
Shlomo Berkovsky	CSIRO, Australia
Albert Bifet	University of Waikato, New Zealand
Robin Burke	DePaul University, USA
Polo Chau	Georgia Institute of Technology, USA
Guanling Chen	University of Massachusetts Lowell, USA
Stephan Doerfel	University of Kassel, Germany
Jill Freyne	CSIRO, Australia
Ruth Garcia Gavilanes	Universitat Pompeu de Fabra and Telefonica I+D, Spain
Daniel Gayo-Avello	University of Oviedo, Spain
Eduardo Graells-Garrido	Telefonica I+D, Spain
Michael Granitzer	Universität Passau, Germany
Bin Guo	Institut Telecom SudParis, France
Ido Guy	eBay, Israel
Chi Harold Liu	Beijing Institute of Technology, China
Eelco Herder	L3S Research Center, Germany
Andreas Hotho	Universität Würzburg, Germany
Geert-Jan Houben	Delft University of Technology, The Netherlands
Andreas Kaltenbrunner	Barcelona Media, Spain
Mark Kibanov	University of Kassel, Germany
Bart Knijnenburg	University of California, USA
Dominik Kowald	Graz University of Technology, Austria
Florian Lemmerich	Universität of Würzburg, Germany
Javier L. Canovas Izquierdo	Internet Interdisciplinary Institute (IN3) - UOC, Spain
Leandro Marinho	Universidade Federal de Campina Grande, Brazil
Claudia Müller	Freie Universität Berlin, Germany
Kjetil Norvag	Norwegian University of Technology, Norway
John O'Donovan	University of California, Santa Barbara, USA
Denis Parra	Pontificia Universidad Católica de Chile, Chile
Nico Piatkowski	TU Dortmund University, Germany
Haggai Roitman	IBM Research Haifa, Israel
Shaghayegh Sahebi	University of Pittsburgh, USA
Alan Said	University of Gothenburg, Sweden
Christoph Scholz	University of Kassel, Germany

Philipp Singer	Graz University of Technology, Austria
Maarten Van Someren	University of Amsterdam, The Netherlands
Claudia Wagner	GESIS, Germany
Su Yang	Fudan University, China
Shengdong Zhao	National University of Singapore, Singapore
Arkaitz Zubiaga	The University of Warwick, UK

Additional Reviewers

Lukas Eberhard	Graz University of Technology, Austria
Byungkyu Kang	University of California, Santa Barbara, USA
Patrick Kasper	Graz University of Technology, Austria

Contents

Link Classification and Tie Strength Ranking in Online Social Networks with Exogenous Interaction Networks

Mohammed Abufouda[✉] and Katharina Anna Zweig

Computer Science Department, University of Kaiserslautern,
Gottlieb-Daimler-Str. 48, 67663 Kaiserslautern, Germany
{abufouda,zweig}@cs.uni-kl.de

Abstract. Online social networks (OSNs) have become the main medium for connecting people, sharing knowledge and information, and for communication. The social connections between people using these OSNs are formed as virtual links (e.g., friendship and following connections) that connect people. These links are the heart of today's OSNs as they facilitate all of the activities that the members of a social network can do. However, many of these networks suffer from noisy links, i.e., links that do not reflect a real relationship or links that have a low intensity, that change the structure of the network and prevent accurate analysis of these networks. Hence, a process for assessing and ranking the links in a social network is crucial in order to sustain a healthy and real network. Here, we define link assessment as the process of identifying noisy and non-noisy links in a network. In this paper (The work in this paper is based on and is an extension of our previous work [2].), we address the problem of link assessment and link ranking in social networks using external interaction networks. In addition to a friendship social network, additional exogenous interaction networks are utilized to make the assessment process more meaningful. We employed machine learning classifiers for assessing and ranking the links in the social network of interest using the data from exogenous interaction networks. The method was tested with two different datasets, each containing the social network of interest, with the ground truth, along with the exogenous interaction networks. The results show that it is possible to effectively assess the links of a social network using only the structure of a single network of the exogenous interaction networks, and also using the structure of the whole set of exogenous interaction networks. The experiments showed that some classifiers do better than others regarding both link classification and link ranking. The reasons behind that as well as our recommendation about which classifiers to use are presented.

Keywords: Link assessment · Link ranking · Multiple networks · Social network analysis

M. Atzmueller et al. (Eds.): MUSE 2015/MSM 2015/MSM 2016, LNAI 11406, pp. 1–27, 2019.
https://doi.org/10.1007/978-3-030-34407-8_1

1 Introduction

Online social networks (OSNs) have become a vital part of modern life, facilitating the way people get news, communicate with each other, and acquire knowledge and education. Like many other complex networks, online social networks contain noise, i.e., links that do not reflect a real relationship or links that have a low intensity. These noisy links, especially false-positives, change the real structure of the network and decrease the quality of the network. Accordingly, having a network with a lot of noise impedes accurate analysis of these networks [57]. In biology, for example, researchers often base their analysis of protein-protein interaction networks on so-called high-throughput data. This process is highly erroneous, generating up to 50% false-positives and 50% false-negatives [16] and thus introducing noisy links into the constructed protein-protein interaction networks. As a result, assessing how real a link is in these networks is inevitable in order to get a high quality representation of the studied system. Therefore, getting accurate analysis results is hard to attain without an assessment process. Based on that, many researchers have started assessing the quality of these biological networks [12, 20] by assessing the links of these networks. In social networks, the situation is quite similar, as many online social networks experience such noisy relationships. A friend on Facebook, a follower on Twitter, or a connection on LinkedIn does not necessarily represent a real-life friend, a real person, and a contact from your professional work, respectively. A possible reason for the noisy relationships in these OSNs is the low cost of forming a link on online social network platforms, which results in a large number of connections for a member. Another reason for the existence of noisy relationships is the automatic sending of invitations when a member first registers on one of the social network platforms; these invitations may contribute to connecting you with persons you really do not know in real life but whom you have contacted once for any reason. Another example is the follow relationships in the Twitter social network, where it is easy to be followed by a fake account or by a real account whose owner seeks a possible follow back to get more connections.

In this work, we aim at assessing the relationships within a friendship social network (SN) based on the structure of networks related to the friendship social network of interest SN. These networks are called *Exogenous Interaction Networks:* $\mathcal{G} = \{G_1, G_2, \cdots, G_n\}$. We have shown in previous work [1] that exogenous interaction networks influence the tie formation process in the friendship social network; thus, using information from these networks helps to assess the links in the friendship social network. Looking merely at one individual network, in this case the SN, is a rather simplistic abstraction of social interaction, which is not sufficient for understanding its dynamics [9]. Thus, utilizing the interaction networks that affect the structure of the social network is our concern in this work.

To better understand the concept of link assessment using associated interaction networks, let us consider a research center environment real data set that we will use later in the experiments. Figure 1 depicts a visualization of the networks, where the members can socialize online using the Facebook social network SN, which is chosen as the social network of interest to be assessed later. In addition to the Facebook friendship network, the members of the research group have different interactions that affect the structure of their *Facebook* friendship network SN. These exogenous interaction networks $\mathcal{G}$ include:

- Work G_1: Where a link exists between two members if they work/ed in the same department.
- Co-author G_2: Where a link exists between two members if they have co-authored a publication.
- Lunch G_3: Where a link exists between two members if they had lunch together at least one time.
- Leisure G_4: Where a link exists between two members if they have participated in the same leisure activity at least one time.

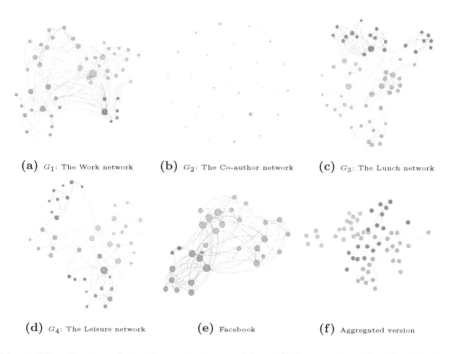

(a) G_1: The Work network   **(b)** G_2: The Co-author network   **(c)** G_3: The Lunch network

(d) G_4: The Leisure network   **(e)** Facebook   **(f)** Aggregated version

Fig. 1. Visualization of the Research Group dataset [32] networks. The visualization was done using Gephi's Yifan Hu [7] visualization algorithm with a few manual edits. The size of the nodes is directly proportional to their degrees and the color of the node is based on the community to which a node belongs [8]. The aggregated version in (f) was obtained by aggregating all of the associated interaction networks into one network: $\bigcup_{i=1}^{4} G_i$, where duplicated edges were removed. (Color figure online)

The interactions in the exogenous networks affect the structure of the social network SN as the link formation process within any social network is not only driven by its structure, i.e., internal homophily [35], but is also influenced extremely by external factors (exogenous interaction networks $\mathcal{G}$) [1]. For example, it is highly probable that any two persons who have had lunch together and/or have spent some leisure time together will be friends in the SN. However, if there is a friendship link between two members A and B in the SN and there is no link between A and B in any of the networks in $\mathcal{G}$, then this relationship might be a noisy one, or it may be a very low strength link that does not qualify as a real friendship relation. In Fig. 2, the links that exist in the social network SN and also exist in any other network $G_i \in \mathcal{G}$ are presumably valid links, as the links in these interaction networks affect the link formation in the social network SN. On the other hand, there are 44 links that are not in any $G_i \in \mathcal{G}$, which leads to the question: *Are these edges noise?* In fact, these 44 links are most likely noisy links. However, it is hard to capture all of the possible relationships between the actors of this dataset in real life. For example, one link of the 44 might be between two researchers who are living in the same building or are members of the same political party, which is data that we do not have or that is hard to collect. Thus, these links are potential noise or relationships with very low intensity.

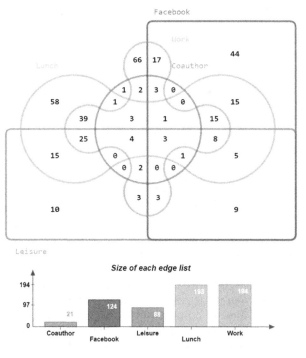

Fig. 2. The Venn diagram [6] for edge overlapping between the Facebook social network SN and the other associated interaction networks $\mathcal{G}$ for the Research Group dataset. (Color figure online)

The reminder of this paper is structured as follows. Section 2 presents the related work. Section 3 contains the definitions and notations used in the paper. Section 4 provides details about the method developed in this work, and Sect. 5 defines the ground truth data set and also explains the experiment setup. Section 6 gives details about the datasets and the evaluation metrics used to evaluate the proposed method. The paper ends with a presentation of the results in Sect. 7, and the Summary in Sect. 9.

2   Related Work

This work can be seen from two different perspectives. The first is *Link prediction* using external information, and second is *Tie strength* ranking. In this section, we will provide the related works based on these two categories and highlight how our work is different.

First, the work in this paper is related to the link prediction problem using external information that is associated with the social networks. Hereafter, we provide the related work.

The problem of link prediction was initially defined in the seminal work of Liben-Nowell and Kleinberg [30] followed by a plethora of research in the area of link prediction. Surveys and literature reviews such as [4, 31, 33, 48] provided an overview of the methods used in link prediction. The most relevant work to our work is link prediction using a social network plus additional information. Wang and Sukthankar [50] provided a link prediction model for predicting the collaboration of the researchers of the DBLP using different types of relations. Yang et al. [53] and Negi and Chaudhury [36] provided link prediction models for multi-relational networks, where the edges have different types of interactions. Similarly, Davis [15] provided a link prediction of Youtube following relationships using different types of interactions that were captured on Youtube such as sharing videos and sharing subscriptions. A similar work was done by Horvat et al. [24] for inferring the structure of a social network using the structures of other social networks. A recent work by Lakshmi and Bhavani [25] incorporated temporal data to the multi-relational dataset to provide an effective link prediction.

The contribution in this paper is different from the aforementioned research as follows. Our method considers not only online activity of the members as social relationships, but also some other offline interactions or interactions that are platform independent, i.e., interactions that took place outside the social network platform. This, gives more insights about the motives behind tie formations in online social networks. Additionally, to the best of our knowledge, the link assessment problem has not been addressed before in the context of social networks as complex systems. Here in this work, we provide a definition of the link assessment problem and a method to quantify the noisy links in them.

Second, the work in this paper is related to the tie strength ranking research. Hereafter, we provide the related work.

A recent study has shown that at least 63% of Facebook users have unfriended at least one friend for different reasons [45]. According to Sibona [45], the reasons for unfriending include frequent or useless posts, political and religious polarization, inappropriate posts, and others. These reasons behind the deletion of a friendship connection mean that social networks suffer from noisy relationships that need to be eliminated in order to keep only the desired friends. Accordingly, it is obvious that online social networks contain many false-positive links that push the members to use the unfriend/unfollow feature or, as a less extreme reaction, categorize unwanted connections as restricted members. Thus, a member of an online social network can easily connect to another member based on strong motivation, like being a real-life friend or participating in the same political party, or based on weak motivation like being a friend of someone they know. This variation in the type of friendship links in social networks has led some researchers to quantify the relationship's strength [18, 19, 21, 26, 51, 55] within social networks. Pappalardo et al. [38] proposed a multidimensional model to capture the strength of the ties in social networks of the same actors. Another related work was done by Xie et al. [52], where the authors studied Twitter's users to identify real friends. Also, Spitz et al. [46] assessed the low-intensity relationships in complex bipartite networks using node-based similarity measures. Pratima and Kaushal [40] provided a prediction model for predicting tie strength between any two connected users of OSNs as an alternative to the binary classification of being a friend or not. Kumar et al. [28] studied the weight prediction, as form of tie strength, in signed networks. Some researchers were interested only in quantifying strong ties. Jones et al. [26] studied the interactions among the users of Facebook to identify the strong ties in the network. A similar recent work by Rotabi et al. [43] employed network motifs to detect strong ties in social networks. Some applications of tie strength have been applied in different domains. Wang [49] et al. provided a social recommendation system based on tie strength prediction. McGee et al. [34] predicted the location of the users using tie strength of the members of Twitter.

The contribution in this paper is different from the aforementioned research as follows. Our methods employed additional information in modeling the tie strength other than the social network. The additional information not only improved the tie strength ranking, but also gives insights regarding what are the external factors that affect the interactions in the social networks. Moreover, we validated the method for link assessment using datasets that include a ground truth which is used also as a gold standard in validating the results of tie ranking among the members of the OSCs.

3   Definitions

An undirected network $G = (V, E)$ is a tuple that is composed of two sets V and E, where V is the set of nodes and E is the set of edges such that an undirected edge e is defined as $e = \{u, v\} \in E$ where $u, v \in V$. For a directed network $\overrightarrow{G} = (V, \overrightarrow{E})$, a directed edge $\overrightarrow{e}$ is defined as $\overrightarrow{e} = (u, v)$ where the node u is the *source* and the node v is the *target*. For undirected networks, the degree of a node w is defined as the number of nodes that are connected to it, while for directed networks the *in-degree* and the *out-degree* are defined as the number of edges in the network where the node w is the target and the source node, respectively.

4 The Proposed Method

This section presents the details of the proposed method. It starts with a general description of the designed framework followed by a detailed information about the feature engineering.

4.1 Framework Description

The aim of this work is to assess and to rank the links in a social network SN using the exogenous interaction networks of the same members of the SN. The proposed method benefits from the structure of these networks in order to infer with the help of a supervised machine learning classifier whether a link in the SN is a true-positive or a false-positive. The idea of the proposed framework is to convert the link assessment problem into a machine learning classification problem. In the following, a description of the framework will be provided.

4.2 Feature Data Model (FDM)

The feature data model is a model that represents a network structure using topological edge-proximity features. More formally, $\forall\, v, w \in V(G_i)$ where $v \neq w$ and $G_i \in \mathcal{G}$, the feature value $F_j(v, w)$ is calculated such that $F_j \in \mathcal{F}$, and $\mathcal{F}$ is the set of features that will be described in the following section.

Edge Proximity Features: The edge proximity features are based on the following measures:

- *The number of Common Neighbors* $(\mathcal{CN})$: For any node z in a network G, the neighbors of z, $\Lambda(z)$, is the set of nodes that are adjacent to z. For each pair of nodes v and w, the number of common neighbors of these two nodes is the number of nodes that are adjacent to both nodes v and w.

$$\mathcal{CN}(v, w) = |\Lambda(v) \cap \Lambda(w)| \tag{1}$$

- *Resource Allocation* $(\mathcal{RA})$: Zhou et al. [56] proposed this measure for addressing link prediction and showed that it provided slightly better performance than $\mathcal{CN}$. This measure assumes that each node has given some resources that will be distributed equally among its neighbors. Then, this idea is adapted by incorporating two nodes v and w.

$$\mathcal{RA}(v, w) = \sum_{\substack{z \in \{\Lambda(v) \cap \Lambda(w)\} \\ z \neq v \neq w}} \frac{1}{|\Lambda(z)|} \tag{2}$$

- *Adamic-Adar Coefficient* $(\mathcal{AAC})$: Ever since this measure was proposed by Adamic et al. [3], the Adamic-Adar Coefficient has been used in different areas of social network analysis, such as link prediction. The idea behind this measure is to count the common neighbors weighted by the inverse of the logarithm.

$$\mathcal{AAC}(v, w) = \sum_{\substack{z \in \{\Lambda(v) \cap \Lambda(w)\} \\ z \neq v \neq w}} \frac{1}{log|\Lambda(z)|} \tag{3}$$

– *Jaccard Index* ($\mathcal{JI}$): This measure was first proposed in information retrieval [44] as a method for quantifying the similarity between the contents of two sets. This idea is applied to the neighbors of any two nodes as follows:

$$\mathcal{JI}(v, w) = \frac{|\Lambda(v) \cap \Lambda(w)|}{|\Lambda(v) \cup \Lambda(w)|} \tag{4}$$

– *Preferential Attachment* ($\mathcal{PA}$): Newman [37] showed that in collaboration networks the probability of collaboration between any two nodes (authors) v and w is correlated to the product of $\Lambda(v)$ and $\Lambda(w)$.

$$\mathcal{PA}(v, w) = |\Lambda(v)| \cdot |\Lambda(w)| \tag{5}$$

– *Sørensen-Dice Index* ($\mathcal{SD}$): This measure has been used in ecology to find the similarity between species in ecological data [17] and it is defined as:

$$\mathcal{SD}(v, w) = \frac{2 \times |\Lambda(v) \cap \Lambda(w)|}{|\Lambda(v)| + |\Lambda(w)|} \tag{6}$$

– *Hub Promoted Index* ($\mathcal{HPI}$): This measure was used to find the similarity between two nodes in a networks with hierarchical structures [42], and it is defined as:

$$\mathcal{HPI}(v, w) = \frac{|\Lambda(v) \cap \Lambda(w)|}{min(|\Lambda(v)|, |\Lambda(w)|)} \tag{7}$$

– *Hub Depressed Index* ($\mathcal{HDI}$): Similar to $\mathcal{HPI}$, the $\mathcal{HDI}$ is defined as:

$$\mathcal{HDI}(v, w) = \frac{|\Lambda(v) \cap \Lambda(w)|}{max(|\Lambda(v)|, |\Lambda(w)|)} \tag{8}$$

– *Local community degree measures* ($\mathcal{CAR}$)[1]: Measuring the similarity between two nodes can also be done by looking at how the common neighbors of these two nodes are connected to these two nodes. The common neighbor measure based on the local community degree measure was introduced in [10] and is defined as:

$$\mathcal{CAR}(v, w) = \sum_{\substack{z \in \{\Lambda(v) \cap \Lambda(w)\} \\ z \neq v \neq w}} \frac{|\Lambda(v) \cap \Lambda(w) \cap \Lambda(z)|}{|\Lambda(z)|} \tag{9}$$

The similarity measures described above are used for undirected networks. For directed networks, two versions of each measure are used by providing two versions of the neighborhood set Λ, the in-neighbors $\Lambda(v)_{in}$ and the out-neighbors $\Lambda(v)_{out}$. Based on this, an *in* and an *out* version of the above measures can be constructed. For example, the $\mathcal{CN}_{in}$ for two nodes v, w is: $\mathcal{CN}(v, w)_{in} = |\Lambda(v)_{in} \cap \Lambda(w)_{in}|$.

---

[1] We stick to the name $\mathcal{CAR}$ as provided by the authors in [10].

Network Global Features: Assume that FDM_{G_i} is the feature data model constructed from the network G_i, then we call $FDM_{\mathcal{G}}$ the aggregated model from all networks $\mathcal{G}$. For $FDM_{\mathcal{G}}$, network global features are required to represent the global properties of each network G_i. This means that a pair of nodes v, w appears $|\mathcal{G}|$ times in the combined $FDM_{\mathcal{G}}$. These global features help the classification algorithm discriminate among different instances of v, w if their edge proximity features are close to each other. Therefore, it is crucial to label the instances in the $FDM_{\mathcal{G}}$ with network global features. Network density is used as a network global feature. *Network density* (η): Is a measure that reflects the degree of completeness of a network, and it is defined as:

$$\eta(G_i) = \frac{2 \cdot |E(G_i)|}{|V(G_i)| \cdot (|V(G_i)| - 1)} \tag{10}$$

Based on the above description, the $FDM_{\mathcal{G}}$ for an undirected network contains a number of instances that is equal to $\sum_{G_i \in \mathcal{G}} \frac{|V(G_i)| \cdot (|V(G_i)| - 1)}{2}$ such that for every pair of nodes $v, w \in V(G_i)$ and $G_i \in \mathcal{G}$, an instance $\mathcal{I}(v, w)$ is a tuple that contains: (1) the edge proximity features' values for v and w presented in Eqs. 1 to 9; (2) the network global feature of $\mathcal{G}$ presented in Eq. 10; (3) a binary class, $\{1, 0\}$, which indicates whether there is a link $e = \{v, w\}$ in G_i or not. This binary class is what we are predicting here.

Figure 3 depicts the process of assessing the links of a social network SN using associated interaction networks $\mathcal{G}$. In step 1, the FDM_{G_i} is constructed for each network $G_i \in \mathcal{G}$. In step 2, the constructed FDMs are used to train a machine learning classifier which is used, in step 3, to assess the SN by providing the binary classification value. In step 4, the ground truth labels are used to evaluate the classification performance. Based on this method, training and testing are done on two disjoint sets, except when training and testing on the SN, which enhances and supports the results, as we will see later in Sect. 7.

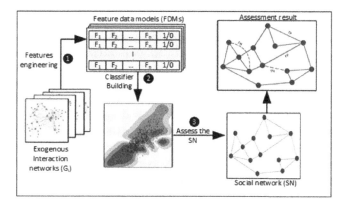

Fig. 3. The framework for link assessment using associated interaction networks and machine learning. (Color figure online)

5   Ground Truth and Experiment Setup

Let the ground truth $SN = (V, E)$ be the network with the set of nodes V and with the edge set E that contains only true-positives and true-negatives. Let $SN_{predicted} = (V, E')$ be the predicted social network on the same set of nodes V and the set of edges E' is the predicted edges. Accordingly, the set $E - E'$ contains the false-negative links, i.e., links that exist in reality (in E) but were not observed in the $SN_{predicted}$ (in E'). Similarly, $E' - E$ contains the false-positive links, i.e., those that do not exist in SN but were observed in the $SN_{predicted}$. The goal is now to get a classification result that is as close to SN as possible. Therefore, the more accurate the machine learning classifier, the more efficient the link assessment method.

Based on that, the machine learning problem $\psi(X, Y)$ means that the data X is used to train a machine learning classifier to classify the links in Y, where $X \neq Y$. To test the effectiveness of this method, a social network with ground truth data will be assessed. If the links of the social network SN are assessed using one network $G_i \in \mathcal{G}$, then the machine learning problem becomes: $\psi(FDM_{G_i}, SN)$, which means that the training phase uses the FDM generated only from a single network G_i to assess the links in (to test on) the SN. In this case, global network feature is excluded, as it is fixed for all instances of the same network, and thus is useless for the classifier algorithm. This assessment enables us to determine whether the structure of a network $G_i \in \mathcal{G}$ is sufficient to efficiently assess the links in the SN or not. Additionally, this will provide insights regarding the correlation between this single network and the social network. Similarly, if the links of the SN are assessed using the whole set of the interaction networks, then the machine learning problem becomes: $\psi(FDM_{\mathcal{G}}, SN)$, which means that the training phase uses the aggregated FDMs of all interaction networks, and in this case the global network feature is included.

In order to test the proposed solution, the following experiment steps were performed:

1. *Build the FDM*: In this step the values of the features described in Sect. 4.2 were calculated, which constructs the FDM_{G_i} for every network G_i of the interaction networks $\mathcal{G}$, and also for the social network of interest FDM_{SN}, where SN is the ground truth to test on.
2. *Training and testing*: The classifier was trained using different training sets depending on the goal of the experiment. To assess the links of the social network of interest SN using a single network $G_i \in \mathcal{G}$, the training set was FDM_{G_i} and the test set was the FDM_{SN}. For assessing the links of the social network of interest SN using the whole set of the associated interaction networks, the training set was $FDM_{\mathcal{G}}$ and the test set was the FDM_{SN}. To assess the links of the social network of interest SN using the SN itself, training and testing were done using the FDM_{SN} with k-fold cross-validation.
3. *Evaluation*: The evaluation metrics described in Sect. 6.2 were used to evaluate the classification results of the training sets.

6   Datasets and Evaluation Metrics

In this section, a description of the datasets and the evaluation metrics will be presented.

6.1   Datasets

In order to validate the proposed method, we tested it using two different social networks with their associated interaction networks. The first dataset (RG) was the research group dataset described in Sect. 1. The social network for the research group [32], which is the Facebook social network, was considered as the ground truth online social network for its members[2]. All the networks of the first dataset are undirected. The second dataset (LF) was a *law firm* dataset [29] containing an offline directed social network along with the following two exogenous interaction networks based on questionnaires:

– *Advice* G_1: If a member A seeks advice from another member B, then there is a directed link from A to B.
– *Cowork* G_2: If a member A considers another member B a co-worker, then there is a directed link from A to B.

The *Friendship* network of the law firm dataset and the *Facebook* social network of the research group dataset are considered as the ground truth social networks. That is because both networks were validated by the collectors and the edges in both networks are true-positives and the edges that are absent are true-negatives[3]. Table 1 shows the network statistics of the networks of the datasets used in this paper. These statistics include the number of nodes n, the number of links m, the average clustering coefficient $cc(G)$, and the network's density η.

Table 1. Datasets statistics.

Dataset	Networks	n	m	$cc(G_i)$	$\eta(G_i)$
RG	SN: **Facebook**	32	248	0.48	0.5
	G_1: Work	60	338	0.34	0.19
	G_2: Co-author	25	42	0.43	0.14
	G_3: Lunch	60	386	0.57	0.21
	G_4: Leisure	47	176	0.34	0.16
LF	SN: **Friends**	69	339	0.43	0.07
	G_1: Co-work	71	726	0.41	0.15
	G_2: Advice	71	717	0.42	0.14

---

[2] This may sound contradictory to what we claimed in the introduction concerning noise in social networks. However, we contacted the owner of the dataset and made sure that there are neither false-positives nor false-negatives in the Facebook network.

[3] A description of the law firm dataset and how it was collected can be found on the original publisher page.

6.2 Evaluation Metrics

In order to evaluate the prediction results, we present a set of classical classification evaluation metrics used for evaluating the classification results of the experiment. For any two nodes v and w, a true-positive (TP) classification instance means that there is a link $e = \{v, w\}$ between these two nodes in the test set, for example the links of the SN, and the classifier succeeds in predicting this link. A true-negative (TN) instance means that there is no link $e = \{v, w\}$ in the test set and the classifier predicts that this link does not exist. On the other hand, a false-negative (FN) instance means that for a pair of nodes v and w there is a link $e = \{v, w\}$ in the test set and the classifier predicts that there is no link. Similarly, false-positive (FP) instance means that for a pair of nodes v and w there is no link $e = \{v, w\}$ in the test set but the classifier predicts that there is a link.

Based on the basic metrics described above, we used the following additional evaluation metrics:

- **Precision** ($\mathcal{P}$): the number of true-positives in relation to all positive classifications. It is defined as: $\mathcal{P} = \frac{TP}{TP+FP}$
- **Recall** ($\mathcal{R}$): also called True Positive rate or *Sensitivity*. It is defined as: $\mathcal{R} = \frac{TP}{TP+FN}$
- **Accuracy** ($\mathcal{ACC}$): the percentage of correctly classified instances. It is defined as: $\mathcal{ACC} = \frac{TP+TN}{TP+TN+FP+FN}$
- **F-measure** ($\mathcal{F}$): the harmonic mean of precision and recall. It is defined as: $\mathcal{F} = \frac{2 \cdot \mathcal{P} \cdot \mathcal{R}}{\mathcal{P}+\mathcal{R}}$

Those measures, particularly the accuracy, are not informative if there are imbalanced datasets where one class, the no-edge class in the FDM, comprises the majority of the dataset instances. To achieve more rigorous validation of the results, we used the weighted version of the above measures to reflect on informative and accurate measures.

Area Under Receiver Operating Characteristics Curve ($AU\text{-}ROC$): The ROC [22] curve plots the true positive rate against the false positive rate. The area under this curve reflects how good a classifier is and is used to compare the performance of multiple classifiers.

7 Results

In this section, the properties of the constructed $FDMs$ and the classification results will be presented.

7.1 The Properties of the FDM

In this section, some properties of the FDM's feature and what they look like will be presented. Figure 4 shows a selected two dimensions(2-D) of the FDMs constructed from the used networks. The figure shows that the FDM is not linearly separable, which renders the classification problem non-trivial for linear classification models. The figure shows also that there are some features that are highly correlated, for example, Fig. 4h shows a strong correlation between the $\mathcal{SD}$ and the $\mathcal{HDI}$ features. Later, we discuss the correlation between the features and their impact on the classification process.

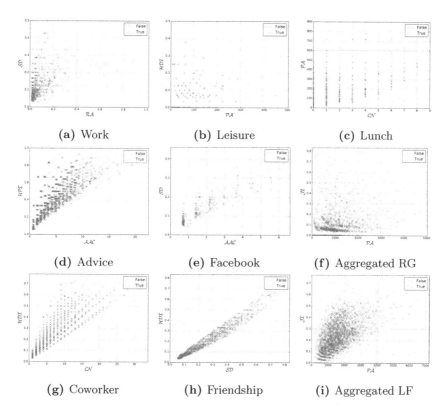

(a) Work   **(b)** Leisure   **(c)** Lunch

(d) Advice   **(e)** Facebook   **(f)** Aggregated RG

(g) Coworker   **(h)** Friendship   **(i)** Aggregated LF

Fig. 4. Selected 2-D scatters of the FDM for the used networks. The x-axis and the y-axis represent selected features presented in Eqs. 1 to 9. The red markers are the *False* instances and green markers are the *True* instances which indicate the existence and the non-existence of an edge, respectively. (Color figure online)

There are many machine learning classifiers, each with its own assumptions, limitations, and parameters to tune. For example, some classifiers like *Logistic Regression* assumes that there is no correlation between the features. This makes logistic regression not suited for classification with correlated features. Whereas, there are classifiers, such as *Support Vector Classifier* with kernels, which can perform well with correlated features; others assume a Gaussian distribution of the features, and so on. Thus, it is crucial to understand the data that is being used in the classification process. Figures 5 and 6 shows deeper analysis of the FDM's features. In Fig. 5, the correlation between the features of the FDM is not the same across all networks of the *Research Group* dataset. In Fig. 5a (the Work network), there is less correlation between the features when compared with, for example, Fig. 5i (Facebook). From Fig. 5, panels 5a, c, e, g, i, and k, the feature that is correlated the least with the other features is the $\mathcal{PA}$. It turned out that the FDM's features are intrinsically correlated. The reason is that, unlike the other features,

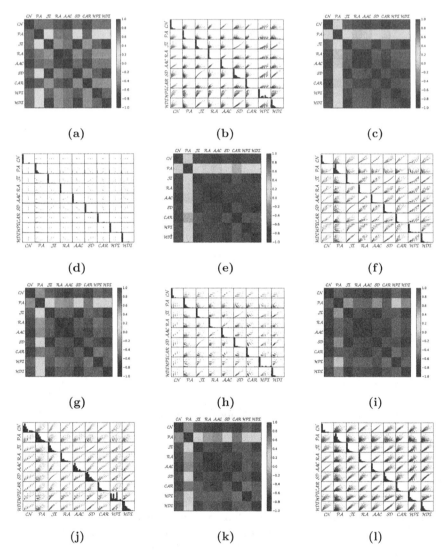

Fig. 5. The feature correlation matrix and the feature correlation scatter of the features of the FDM for the Research Group (RG) dataset. Panels a, c, e, g, i, and k show the correlation matrix for the FDM of the networks Work, Coauthor, Lunch, Leisure, Facebook, and Aggregated RG, respectively. Panels b, d, f, h, j, and l show the correlation scatter between two feature of the FDM for those networks, with the distribution of each feature in the diagonal. (Color figure online)

the $\mathcal{PA}$ feature is not dependent in the $\mathcal{CN}$. The correlation is clearer in the corresponding correlation scatters in Fig. 5, panels 5b, d, f, h, j, and l. These panels show a strong correlation between $\mathcal{JI}$ and $\mathcal{SD}$, between $\mathcal{JI}$ and $\mathcal{HDI}$, and between $\mathcal{ACC}$ and $\mathcal{CN}$. Also, from the distribution of the feature in the diagonals of Fig. 5, panels 5b, d, f, h, j, and l, it is obvious that the distribution of these features is not Gaussian. Most features of all $FDMs$ show low variance, except for the FDM of Facebook in Figs. 5i and j. This will affect the performance of the classifiers as we will see later.

Figure 6 shows the same analysis as presented in Fig. 5 but for the *Law Firm* dataset. However, there are some differences in the properties of the features of the $FDMs$ of the Law Firm networks. For example, the networks' $FDMs$ have more variance for all features of the $FDMs$ of all networks. Also, the features are more correlated with each other when compared to the Research Group dataset.

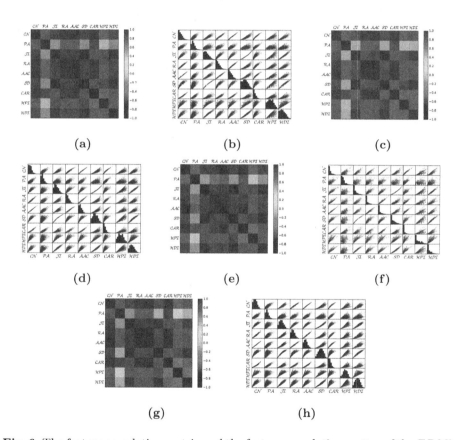

Fig. 6. The feature correlation matrix and the feature correlation scatter of the FDM's feature for the Law Firm (LF) dataset. Panels a, c, e, and g show the correlation matrix for the FDM of the networks Advice, Coworker, Friend, and Aggregated LF, respectively. Panels b, d, f, and h show the correlation scatter between two features each of the FDM for those networks, with the distribution of each feature in the diagonal. (Color figure online)

7.2   Classification Results

Next, we will present the results of the method proposed in this work. The results were first obtained for random graphs as a null model. The results of the null model were insignificant compared to the results presented here [2]. The results presented in this sections are based on the $\mathcal{SVM}$ classifier [13] with a Gaussian kernel. Table 2 shows the results of the assessment for the Research Group dataset. The assessment results are satisfactory in terms of the evaluation metrics. The lower bound for the classification is 0.824 in terms of the F-measure, which is good considering the very small data the FDM of the network Coauthor contains (cf. Fig. 1b to see how small this network is). Note that this lower bound was improved compared to our previous work in [1] due to incorporating additional edge proximity measures. Surprisingly, the results in the table also show that the aggregated FDM does not provide any noticeable advantage over the single networks. Having said that, the Lunch and the Work networks provided the best results, which suggests the friendship in the Research Group dataset is highly correlated with the Lunch network, which seems reasonable as we tend to have lunch often with our friends, but it is not necessarily that we coauthor with a friend.

Table 2. The prediction results for the Research Group dataset. Note that k-fold cross-validation was used when training and testing on SN.

Dataset	Train on	Test on	Performance			
			$\mathcal{ACC}$	$\mathcal{P}$	$\mathcal{R}$	$\mathcal{F}$
RG	G_1: Work	SN: Facebook	0.841	0.842	0.841	0.841
	G_2: Coauthor		0.822	0.827	0.822	0.824
	G_3: Lunch		0.843	0.835	0.843	0.839
	G_4: Leisure		0.837	0.835	0.836	0.836
	Aggregated		0.834	0.834	0.830	0.836
	SN: Facebook		0.833	0.829	0.830	0.832

Table 3 shows the classification results for the Law Firm dataset. The results in the table shows better performance compared to the Research Group dataset with a lower bound of 0.88 for the F-measure.

We think that the slight advantage in the performance of the Law Firm dataset over the Research Group dataset is due to the higher variance in the FDM's features of the Law Firm as shown in Fig. 6. Even though the RG dataset has more networks, its aggregated FDM did not show better results than the single networks. This indicates that for a better classification of the links, we need more features that capture the structure of the network, other than the edge proximity features that we used. In addition, the results indicate that directed networks may contain more patterns regarding the interaction among the members of these networks.

Table 3. The prediction results for the Law Firm dataset. Note that k-fold cross-validation was used when training and testing on SN.

Dataset	Train on	Test on	Performance			
			$\mathcal{ACC}$	$\mathcal{P}$	$\mathcal{R}$	$\mathcal{F}$
LF	G1: Cowork	SN: Friend	0.889	0.884	0.889	0.886
	G2: Advice		0.893	0.887	0.893	0.889
	Aggregated		0.885	0.879	0.885	0.881
	SN: Friend		0.972	0.984	0.919	0.947

7.3 Comparing Different Classifiers

There are dozens of machine learning classifiers, and each has its advantages, limitations, and parameters to tune, which makes the selection of the appropriate classifier a difficult task. Table 4 shows a comparison of the performance of different classifiers. Based on the results in the table, the presented method showed close performance for most classifiers. Once again, the results of the LF dataset are slightly better than those of the RG dataset for all of the compared classifiers.

Table 4. Comparison of the performance of different classifiers for the aggregated versions of the RG and the LF datasets. The compared classifiers are: The $\mathcal{KN}$: k-Nearest Neighbors vote [5]; $\mathcal{SVM}$: Support Vector Machines [13]; $\mathcal{DT}$: decision trees [41]; $\mathcal{NB}$: Naive Bayes [54]; $\mathcal{LR}$: Logistic Regression [47]. We used the *scikit-learn* Python package [39].

Dataset	Classifier	Performance			
		$\mathcal{ACC}$	$\mathcal{P}$	$\mathcal{R}$	$\mathcal{F}$
RG	$\mathcal{KN}$	0.800	0.795	0.800	0.765
	$\mathcal{SVM}$	0.821	0.833	0.821	0.825
	$\mathcal{DT}$	0.800	0.806	0.800	0.804
	$\mathcal{NB}$	0.778	0.821	0.778	0.780
	$\mathcal{LR}$	0.827	0.825	0.827	0.827
LF	$\mathcal{KN}$	0.843	0.823	0.843	0.794
	$\mathcal{SVM}$	0.816	0.858	0.816	0.830
	$\mathcal{DT}$	0.880	0.870	0.880	0.870
	$\mathcal{NB}$	0.883	0.875	0.883	0.877
	$\mathcal{LR}$	0.868	0.878	0.868	0.872

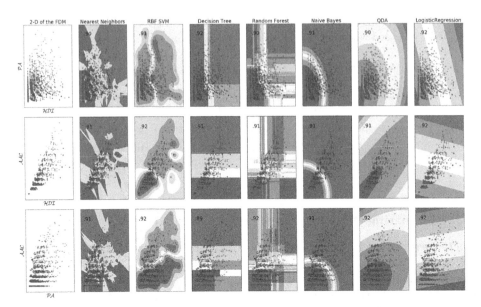

Fig. 7. The decision boundaries for different probability-based classifiers. We used 2-d scatter of the FDM constructed from the aggregated networks of the RG dataset as an illustration. The leftmost panels are the 2-d features before the classification was performed. The red points are False instances, the blue points are True instances. The red "+" markers and the blue "+" markers are the False and the True instances to be classified by the classifier, i.e., the test samples. The other points, none "+" points, are the training points, where the training and the testing points were randomly selected with ratio 75:25, respectively. The other panels represent the classification results with the decision boundaries. The number in the top-left is the accuracy of the classification, and the gradient of the colored areas represents the probability. For example, the darker the blue area, the higher the probability that the points in this area are true. The classifiers used are those classifiers that give a probability as a classification result, and they are, namely: $\mathcal{KN}$: k-Nearest Neighbors vote [5]; $\mathcal{SVM}$ with Gaussian kernel [13]; $\mathcal{DT}$: decision trees [41]; Random Forests [23]; $\mathcal{NB}$: Naive Bayes [54]; $\mathcal{QDA}$: the Quadratic Discriminant Analysis [14]; $\mathcal{LR}$: Logistic Regression [47]. (Color figure online)

Another aspect that is important when talking about different classifiers is the resulting decision boundaries and how good they are. Figure 7 shows the decision boundaries for different classifiers. The figure shows that linear models, like linear $\mathcal{DT}$ and $\mathcal{LR}$, were not able to really discriminate between the False and the True instances efficiently. Additionally, the figure shows that the accuracy metric is a useless measure as it is not informative for the case of the FDM, whose labels are highly imbalanced. For example, let us take a closer look at the $\mathcal{QDA}$ classifier for the second panel, the attributes $\mathcal{HDI}$ vs $\mathcal{AAC}$. The accuracy of the classifier is 0.91 which is considered high. Having said that, the panel shows that *all* of the points were classified in the red area, which ignores the True instances and make it hard to find a binary threshold to produce binary results. Such a behavior indicates that the accuracy is not a good measure to use if we have imbalanced data. On the other hand, classifiers that use kernels (a method to transferring the non-linearly separable data into linearly separable data by transforming the data into a higher dimension) showed good discrimination between the False and the True instances. An example of this is $\mathcal{SVM}$ with Gaussian kernel [13], the third column in Fig. 7. From the figure, it is clear that the $\mathcal{SVM}$ with Gaussian kernel is able to find disjoint areas for the data points, which helps in producing good classification results.

Another way to compare the performance of different classifiers is to use the area under the ROC curve. Figure 8 shows the AUC for $\mathcal{SVM}$ and $\mathcal{LR}$ with differ-ent tuning parameters. Figure 8a again shows that the linear models are not robust and are not able to provide a good classification. Figures 8, panels 8 b, c, d, e, f, g, and h show that $\mathcal{SVM}$ with Gaussian kernel provided a stable performance, which is why we used it for the results presented in Sect. 7.2.

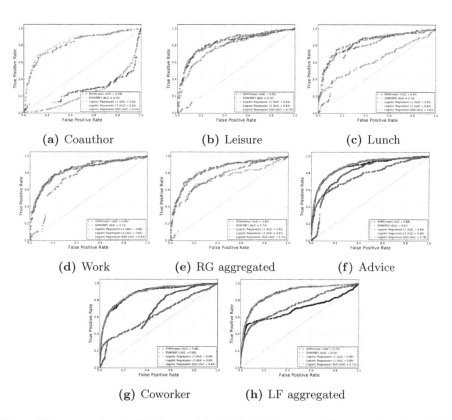

Fig. 8. The area under the ROC curve for $\mathcal{SVM}$ with linear and Gaussian kernels and for $\mathcal{LR}$ with $L1$ and $L2$ regularization and with a stochastic gradient descent optimization algorithm. (Color figure online)

7.4   From Binary Classification to Tie Strength Ranking

In some scenarios, the links of a social network need to be ranked by the tie strength between the members. The proposed method can also give a continuous range of value between 0 and 1, instead of having two classes, using probabilistic classifiers: classifiers that produce a probability value instead of a binary class then finding a threshold for binarizing the resulted probabilities. These probabilities are used here as a *tie strength rank* of the edges in the social network being assessed. Figure 9 shows the ranking results of the SN in the RG and the LF datasets using different classifiers. Our assumption here is that the *best* ranking for the edges of the SN is a step function that changes the values of the edge from *zero* to *one* on the number

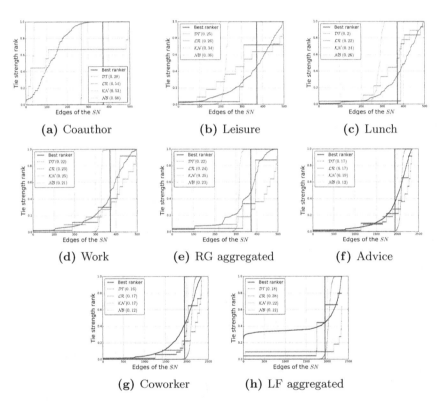

Fig. 9. The Tie strength ranking for the social network using the associated networks. The x-axis represents the edges in the SN ranked by their strength according to the ranking results; the y-axis is the tie strength rank. The best ranker, in bold red, is simply the step function on the number of edges in the social network. The best ranker is used to compare the goodness of the ranking using the proposed method. In the legend, different classifiers are used to estimate the probabilities. The numbers beside the names of the classifiers are the errors in the ranking. This error is calculated as: $\sum_{e \in E_{SN}} |e_{obs} - e_{real}|$, where E_{SN} is the set of edges being ranked in the SN, $e_{obs} \in [0, 1]$ is the probability of having an edge e in the SN, and $e_{real} \in \{1, 0\}$ which means whether e is a true-positive or a true-negative, respectively. (Color figure online)

of true-negative edges in the ground truth network. That means, for undirected SN with n nodes and m edges we have m edges with tie strength *one* and $\binom{n}{2} - m$ edges with tie strength *zero*. Then, the predicted tie strength for all edges, including the true-negatives in the ground truth, is compared to the best ranking using the following error measure: $\sum_{e \in E_{SN}} |e_{obs} - e_{real}|$, where E_{SN} is the set of edges being ranked, including true-negatives, in the SN, $e_{obs} \in [0, 1]$ is the probability of having an edge e in the SN, and $e_{real} \in \{1, 0\}$ which means whether e is a true-positive or a true-negative, respectively. The closer the results to the step function, the better the ranking (cf. Fig. 9).

Figure 9 shows the results of the ranking using the proposed method. $\mathcal{NB}$ and $\mathcal{KN}$ provided the best ranking among all of the classifiers we used. For the RG dataset, the best link ranking, in terms of error ranking as explained earlier, was achieved using the Lunch network with error 20% and by the Work network with error 21% using the $\mathcal{DT}$ and $\mathcal{NB}$, respectively. For the LF dataset, the best ranking was achieved using the any of the networks in the dataset with 12% error using the $\mathcal{NB}$. As in the assessment results presented in the previous section, the ranking results of the SN of the LF is better than the ranking of the RG's SN. Again, we think that the directed networks embrace more information about the structure and the relationships among their members.

7.5   Noisy-Edges Identification

The results in Sect. 7.2 showed how the proposed method is only good in finding true-positive and true-negative edges in a social network. The datasets used in this work do not contain any noisy-edges in the social network, which does not allow a proper validation for noise (false-positives) identification in their original form. Thus, we injected noisy-edges in the SN and tested the method to find how good it is in finding them. To do so, we added k edges to the SN such that $k = \lfloor (\binom{n}{2} - m) \times r \rfloor$ edges, where m is the number of edges in the SN, i.e., the true-positives, and r is the percentage of edges to be added. The resulted network is called $SN_{disguised}$. For example, if $r = 1$ then the $SN_{disguised}$ is a complete network. Then, we predicted only these k edges using the method. The *success rate* is defined as the number of edges which were predicted as false-positive divided by k. We used different values for r ranging between 0.1 and 1.0. The results of the noise identification came as follows. For the research group dataset, the success rates were $0.34, 0.94, 0.94, 0.95$, and 0.97 when training on Coauthor, Leisure, Lunch, Work, and the aggregated version, respectively, and testing only the k edges in $SN_{disguised}$. As the edges to be added to the SN were randomly selected, the results were obtained as the mean of 10 runs for each r averaged by the number of values used of r. The poor performance for the coauthor network is because it is very small network, and it hardly captures a good structure for the relationships among the members. For the law firm dataset, the success rate was 0.99, for all networks with the same settings of r as in the research group dataset. The noise identification in the law firm dataset was higher than the research group dataset. However, the success rate for the aggregated networks of the research group showed a better performance than the performance of the best network, the work network.

8   Discussion

The proposed method showed a good potential in both link classification and tie strength ranking. It seems that machine learning can effectively be used for the network based features. In this section we provide our final thoughts about the problem addressed in the paper, the used method, and the limitations.

8.1   The Importance of Link Assessment and Tie Ranking

Addressing the link assessment problem is crucial in today's life, where online social media contain a lot of spam, ads-intensive websites, and inaccurate news. We strongly think that identifying noisy links in social networks contributes in eliminating these problems and reducing their impacts. Tie strength ranking, on the other hand, can also improve the quality of information spread in online social networks. For example, automatic ranking of the friends list on Facebook might led to better news feed, friends recommendations, and better targeted ads, just to name a few. Thus, the work presented in this paper has actionable insights on online social networks.

It seems that the tie strength problem explicitly includes link assessment. However, the two problems should be separately handled because the cost of link assessment may be lower than the cost of tie strength ranking. One reason for that is the hardness of getting the ground truth of real tie strengths between the nodes of a network. This reason pushed some researchers to concentrate only on the strong ties, like the work presented in the related work section. Another reason is that, the link classification is sufficient in certain applications such as spam detection. Moreover, link classification can be a preprocessing step for many network based analysis tasks, such as community detection, where eliminating noise edges may provide more meaningful communities. Thus, we emphasis the distinction between the two problems.

8.2   Classification Methods for Network-Based Features

Features Correlation: The major network-based features for link proximity are based on common neighbors $\mathcal{CN}$, which makes most of the features highly correlated to each other. Having said that, highly correlated features might be a problem in classification, especially with small number of training data. Thus, devising new link proximity measures that are not based on the number of common neighbors is important.

Data Imbalance: Imbalanced dataset is a common challenge in classification problems. In social networks case, this problem is more vivid because most social networks are sparse. Thus, any edge proximity-based model is inherently imbalanced. Many techniques exists in the literature to avoid the classification limitation under imbalanced datasets [27]. In this work, we used SMOTE (Synthetic Minority Over-sampling Technique) [11], which did not give any improvement in the prediction performance due to the small datasets we have. The literature contains a lot of techniques that can be used with larger datasets to handle the imbalanced nature of some datasets [27].

Classifier Selection: The decision boundaries are helpful to select a good classifier for the used dataset. Learning and optimization processes are computationally expensive, and experimenting different classifiers with different parameters is always laborious task. Thus, experimenting on sample of the data to select the best classifier is crucial. To handle that, decision boundaries, like what presented in Fig. 7, helps a lot in understanding the data that we have and also to select the best classifier for subsequent optimization. Linear classifiers showed poor performance as the constructed FDM is not linearly separable. Thus, using classifiers with kernels showed better performance. Additionally, $\mathcal{KN}$ and $\mathcal{DT}$ showed promising results for the ranking problem. It turned out that these two classifiers provided good probabilities for approximating the tie strength in the SN, but bad thresholds for the binary classification.

Data cleansing, especially outliers removal, may improve the prediction results. However, in this work we did not remove any data as the used datasets are relatively small. For example, when removing all data that is 3 times the standard deviation away from the mean, the results were not as good as the presented in the results section, though, removing outliers might improve the results in the case of having larger datasets.

Baseline Comparison: To provide more confidence for the results, we conducted the experiments on random graphs as a null model. The results of the random graphs were uncomparable to the results of the real used datasets[4]. Additionally, we tested the model against a classifier that uses one simple rule as a baseline prediction. The results of the presented method using the classifiers presented in Sect. 2 were significantly better than the baseline classifier. Finally, a random classifier was used as another baseline classifier, (cf. Fig. 8). The results of the used classifiers were significantly better than random classifier. Thus, we strongly think that the results provided in this paper are significant and are not due to any random chances.

8.3 Limitations

The presented method used data from a social network itself in addition to external information. The external information may not always be available, which represents a challenge. Moreover, the existence of the ground truth data for the tie strength ranking is hard to attain. Thus, we resort to the binary ranker as a gold standard measure to evaluate the tie ranking provided by the method.

9 Summary

In this paper, we presented a method for link assessment and link ranking of the links of online social networks using external social interaction networks. The proposed method employed machine learning classification techniques to perform the

---

[4] More details about the results of the random graphs can be found in our earlier work [2].

link assessment via label classification based on edge-proximity measures. We have conducted experiments on two different datasets that contain a friendship social network in addition to the external social interactions. The link assessment results, in terms of the F1-score and the accuracy, were satisfactory compared to baseline predictors. The results show that it is possible to assess the links in a social network using external social interactions. Additionally, link ranking has also been performed using probabilistic binary classifiers. The intensive study of the features used in this work and the conducted experiments revealed insights about using machine learning for network-based features. These insights are about (1) features correlation and its effect on the classification; (2) label imbalance handling; (3) goodness of the decision boundaries of the used classifiers; (4) classifier selection for both link assessment and link ranking.

From network perspective, the results of the used datasets suggest that directed networks embrace more building structures that enable better link assessment and link ranking compared to the undirected networks. Also, the results suggest that one external interaction networks embraces enough information to assess or rank the links in the social networks. It seems that for a set of persons, a social interaction outside the social network is enough to know much information about their social relationships. From machine learning perspective, the results achieved in this work, for both link assessment and link ranking, show that network-based features can be used in analyzing networks and building prediction models. Additionally, we discovered that some classifiers are good in providing a binary classification for link assessment, while some others are good in providing a probability range for link ranking.

Future work includes utilizing new features that are not based on common neighbors, implementing techniques for handling imbalanced labels, and incorporating feature selection before using the classifiers on the whole set of features.

References

1. Abufouda, M., Zweig, K.: Interactions around social networks matter: predicting the social network from associated interaction networks. In: IEEE/ACM International Conference on Advances in Social Networks Analysis and Mining, pp. 142–145 (2014)
2. Abufouda, M., Zweig, K.A.: Are we really friends?: Link assessment in social networks using multiple associated interaction networks. In: Proceedings of the 24th International Conference on World Wide Web, WWW 2015 Companion, pp. 771–776. ACM, New York (2015)
3. Adamic, L.A., Adar, E.: Friends and neighbors on the web. Soc. Netw. **25**(3), 211–230 (2003)
4. Al Hasan, M., Zaki, M.J.: A survey of link prediction in social networks. In: Aggarwal, C. (ed.) Social Network Data Analytics, pp. 243–275. Springer, Boston (2011). https://doi.org/10.1007/978-1-4419-8462-3_9
5. Altman, N.S.: An introduction to kernel and nearest-neighbor nonparametric regression. Am. Stat. **46**(3), 175–185 (1992)
6. Bardou, P., Mariette, J., Escudié, F., Djemiel, C., Klopp, C.: jvenn: an interactive Venn diagram viewer. BMC Bioinform. **15**(1), 293 (2014)

7. Bastian, M., Heymann, S., Jacomy, M.: Gephi: an open source software for exploring and manipulating networks (2009)
8. Blondel, V.D., Guillaume, J.-L., Lambiotte, R., Lefebvre, E.: Fast unfolding of communities in large networks. J. Stat. Mech. Theory Exp. **2008**(10), P10008 (2008)
9. Boccaletti, S., et al.: The structure and dynamics of multilayer networks. Phys. Rep. **544**(1), 1–122 (2014)
10. Cannistraci, C.V., Alanis-Lobato, G., Ravasi, T.: From link-prediction in brain connectomes and protein interactomes to the local-community-paradigm in complex networks. Sci. Rep. **3**, 1613 (2013)
11. Chawla, N.V., Bowyer, K.W., Hall, L.O., Kegelmeyer, W.P.: SMOTE: synthetic minority over-sampling technique. J. Artif. Intell. Res. **16**, 321–357 (2002)
12. Chen, J., et al.: Systematic assessment of high-throughput experimental data for reliable protein interactions using network topology. In: 16th IEEE International Conference on Tools with Artificial Intelligence, ICTAI 2004, pp. 368–372. IEEE (2004)
13. Cortes, C., Vapnik, V.: Support-vector networks. Mach. Learn. **20**(3), 273–297 (1995)
14. Cover, T.M.: Geometrical and statistical properties of systems of linear inequalities with applications in pattern recognition. IEEE Trans. Electron. Comput. **14**(3), 326–334 (1965)
15. Davis, D., Lichtenwalter, R., Chawla, N.V.: Multi-relational link prediction in heterogeneous information networks. In: 2011 International Conference on Advances in Social Networks Analysis and Mining, pp. 281–288, July 2011
16. Deane, C.M., et al.: Protein interactions: two methods for assessment of the reliability of high throughput observations. Mol. Cell. Proteomics **1**(5), 349–356 (2002)
17. Dice, L.R.: Measures of the amount of ecologic association between species. Ecology **26**(3), 297–302 (1945)
18. Gilbert, E.: Predicting tie strength in a new medium. In: Proceedings of the ACM 2012 Conference on Computer Supported Cooperative Work, CSCW 2012, pp. 1047–1056. ACM, New York (2012)
19. Gilbert, E., Karahalios, K.: Predicting tie strength with social media. In: Proceedings of the SIGCHI Conference on Human Factors in Computing Systems, pp. 211–220. ACM (2009)
20. Goldberg, D.S., Roth, F.P.: Assessing experimentally derived interactions in a small world. Proc. Natl. Acad. Sci. **100**(8), 4372–4376 (2003)
21. Gupte, M., Eliassi-Rad, T.: Measuring tie strength in implicit social networks. In: Proceedings of the 4th Annual ACM Web Science Conference, WebSci 2012, pp. 109–118. ACM, New York (2012)
22. Hanley, J.A., McNeil, B.J.: The meaning and use of the area under a receiver operating characteristic (ROC) curve. Radiology **143**(1), 29–36 (1982)
23. Ho, T.K.: Random decision forests. In: Proceedings of the Third International Conference on Document Analysis and Recognition, vol. 1, pp. 278–282. IEEE (1995)
24. Horvat, E.-A., Hanselmann, M., Hamprecht, F.A., Zweig, K.A.: One plus one makes three (for social networks). PLOS ONE **7**(4), 1–8 (2012)
25. Jaya Lakshmi, T., Durga Bhavani, S.: Link prediction in temporal heterogeneous networks. In: Wang, G.A., Chau, M., Chen, H. (eds.) PAISI 2017. LNCS, vol. 10241, pp. 83–98. Springer, Cham (2017). https://doi.org/10.1007/978-3-319-57463-9_6
26. Jones, J.J., Settle, J.E., Bond, R.M., Fariss, C.J., Marlow, C., Fowler, J.H.: Inferring tie strength from online directed behavior. PLOS ONE **8**(1), 1–6 (2013)
27. Kotsiantis, S., Kanellopoulos, D., Pintelas, P., et al.: Handling imbalanced datasets: a review. GESTS Int. Trans. Comput. Sci. Eng. **30**(1), 25–36 (2006)

28. Kumar, S., Spezzano, F., Subrahmanian, V.S., Faloutsos, C.: Edge weight prediction in weighted signed networks. In: 2016 IEEE 16th International Conference on Data Mining (ICDM), pp. 221–230, December 2016
29. Lazega, E.: The Collegial Phenomenon: The Social Mechanisms of Cooperation among Peers in a Corporate Law Partnership. Oxford University Press, Oxford (2012)
30. Liben-Nowell, D., Kleinberg, J.: The link-prediction problem for social networks. J. Assoc. Inf. Sci. Technol. **58**(7), 1019–1031 (2007)
31. Lü, L., Zhou, T.: Link prediction in complex networks: a survey. Phys. A Stat. Mech. Appl. **390**(6), 1150–1170 (2011)
32. Magnani, M., Rossi, L.: Formation of multiple networks. In: Greenberg, A.M., Kennedy, W.G., Bos, N.D. (eds.) SBP 2013. LNCS, vol. 7812, pp. 257–264. Springer, Heidelberg (2013). https://doi.org/10.1007/978-3-642-37210-0_28
33. Martínez, V., Berzal, F., Cubero, J.-C.: A survey of link prediction in complex networks. ACM Comput. Surv. (CSUR) **49**(4), 69 (2016)
34. McGee, J., Caverlee, J., Cheng, Z.: Location prediction in social media based on tie strength. In: Proceedings of the 22nd ACM International Conference on Information and Knowledge Management, CIKM 2013, pp. 459–468. ACM, New York (2013)
35. McPherson, M., Smith-Lovin, L., Cook, J.M.: Birds of a feather: homophily in social networks. Annu. Rev. Sociol. **27**, 415–444 (2001)
36. Negi, S., Chaudhury, S.: Link prediction in heterogeneous social networks. In: Proceedings of the 25th ACM International on Conference on Information and Knowledge Management, CIKM 2016, pp. 609–617. ACM, New York (2016)
37. Newman, M.E.: Clustering and preferential attachment in growing networks. Phys. Rev. E **64**(2), 025102 (2001)
38. Pappalardo, L., Rossetti, G., Pedreschi, D.: How well do we know each other? Detecting tie strength in multidimensional social networks. In: 2012 IEEE/ACM International Conference on Advances in Social Networks Analysis and Mining (ASONAM), pp. 1040–1045. IEEE (2012)
39. Pedregosa, F., et al.: Scikit-learn: machine learning in Python. J. Mach. Learn. Res. **12**, 2825–2830 (2011)
40. Pratima, Kaushal, R.: Tie strength prediction in OSN. In: 2016 3rd International Conference on Computing for Sustainable Global Development (INDIACom), pp. 841–844, March 2016
41. Quinlan, J.R.: Induction of decision trees. Mach. Learn. **1**(1), 81–106 (1986)
42. Ravasz, E., Somera, A.L., Mongru, D.A., Oltvai, Z.N., Barabási, A.-L.: Hierarchical organization of modularity in metabolic networks. Science **297**(5586), 1551–1555 (2002)
43. Rotabi, R., Kamath, K., Kleinberg, J., Sharma, A.: Detecting strong ties using network motifs. In: Proceedings of the 26th International Conference on World Wide Web Companion, WWW 2017 Companion, Republic and Canton of Geneva, Switzerland, pp. 983–992. International World Wide Web Conferences Steering Committee (2017)
44. Salton, G., McGill, M.J.: Introduction to Modern Information Retrieval. McGraw-Hill, Inc., New York (1986)
45. Sibona, C.: Unfriending on facebook: context collapse and unfriending behaviors. In: 2014 47th Hawaii International Conference on System Sciences (HICSS), pp. 1676–1685, January 2014
46. Spitz, A., Gimmler, A., Stoeck, T., Zweig, K.A., Horvat, E.-A.: Assessing low-intensity relationships in complex networks. PLOS ONE **11**(4), 1–17 (2016)

47. Walker, S.H., Duncan, D.B.: Estimation of the probability of an event as a function of several independent variables. Biometrika **54**(1–2), 167–179 (1967)
48. Wang, P., Xu, B., Wu, Y., Zhou, X.: Link prediction in social networks: the state-of-the-art. Sci. China Inf. Sci. **58**(1), 1–38 (2015)
49. Wang, X., Lu, W., Ester, M., Wang, C., Chen, C.: Social recommendation with strong and weak ties. In: Proceedings of the 25th ACM International on Conference on Information and Knowledge Management, CIKM 2016, pp. 5–14. ACM, New York (2016)
50. Wang, X., Sukthankar, G.: Link prediction in heterogeneous collaboration networks. In: Missaoui, R., Sarr, I. (eds.) Social Network Analysis - Community Detection and Evolution. LNSN, pp. 165–192. Springer, Cham (2014). https://doi.org/10.1007/978-3-319-12188-8_8
51. Xiang, R., Neville, J., Rogati, M.: Modeling relationship strength in online social networks. In: Proceedings of the 19th International Conference on World Wide Web, WWW 2010, pp. 981–990. ACM, New York (2010)
52. Xie, W., Li, C., Zhu, F., Lim, E.-P., Gong, X.: When a friend in twitter is a friend in life. In: Proceedings of the 4th Annual ACM Web Science Conference, pp. 344–347. ACM (2012)
53. Yang, Y., Chawla, N.V., Sun, Y., Han, J.: Link prediction in heterogeneous networks: influence and time matters. In: Proceedings of the 12th IEEE International Conference on Data Mining, Brussels, Belgium (2012)
54. Zhang, H.: The optimality of Naive Bayes. A A **1**(2), 3 (2004)
55. Zhao, X., et al.: Relationship strength estimation for online social networks with the study on facebook. Neurocomputing **95**, 89–97 (2012)
56. Zhou, T., et al.: Predicting missing links via local information. Eur. Phys. J. B **71**(4), 623–630 (2009)
57. Zweig, K.A.: Network Analysis Literacy: A Practical Approach to Networks Analysis Project Design. Springer, Springer (2014)

Stratification-Oriented Analysis
of Community Structure in Networks
# of Face-to-Face Proximity

Stefan Bloemheuvel, Martin Atzmueller$^{(\boxtimes)}$, and Marie Postma

Tilburg University, Tilburg, The Netherlands
{s.d.bloemheuvel,m.atzmuller,marie.postma}@tilburguniversity.edu

Abstract. Temporal evolution and dynamics of social network interactions provide insights into the formation of social relationships. In this paper, we explored automatic detection of face-to-face proximity during two student meet-ups for the purposes of community detection. The data was collected with the help of wearable sensors. We considered two stratification determinants – time and gender of the participants. Thus, we first examined the structural metrics of the formed networks over time, and also performed an analysis of gender influence on the community structure. Contrary to previous studies, we observed that conversations tended to develop in a parabolic rather than linear manner during both events. Furthermore, the gender attribute showed a considerable effect in community formation.

1 Introduction

Social relationships that are formed during events can be captured in face-to-face contact networks [1,2,6]. Their analysis can provide insights into the dynamics, predictability [54–56] and evolution of communities, see, a.o., [7,37,38]. In the past, the available methods to collect empirical data relied on surveys and diary methods, which are notoriously slow and inaccurate [45]. Recently, novel technologies have been developed that provide new and promising approaches of collecting face-to-face contact data [17,27,63].

In this paper, an adapted and substantially extended revision of [18], we focus on face-to-face proximity as the basic measure of social contact occurrence and investigate the evolving community structure. For data collection, we utilized wearable sensors developed by the SocioPatterns consortium.[1] The proximity tags are based on Radio Frequency Identification technology (RFID chips), capable of detecting close-range and face-to-face proximity (1–1.5 m) with a temporal resolution of 20 s [22]. The most significant advantage of using the SocioPatterns tags is that the human body acts as a radio-frequency blocker at the frequency used on the chips [22]. Therefore, only the signals that are broadcasted directly forward from the person wearing the tag will be detected by other tags. With

---

[1] http://www.sociopatterns.org.

© Springer Nature Switzerland AG 2019
M. Atzmueller et al. (Eds.): MUSE 2015/MSM 2015/MSM 2016, LNAI 11406, pp. 28–43, 2019.
https://doi.org/10.1007/978-3-030-34407-8_2

the help of the potential interlocutors' own bodies, a face-to-face contact can be observed with a probability of over 99% using the interval of 20 s as the minimal contact duration [22].

The goals of our current work were threefold: First, to assess the evolution of contacts of face-to-face proximity, second, to compare the performance of existing algorithms for community detection, i. e., for detecting "densely connected" areas of a network, and third the stratification-oriented analysis using time and gender as a socio-demographic variable. Community detection is rather challenging in that a network can be divisible into several communities according to various criteria, sometimes also with a hierarchical structure [28,47]. Practical applications of such algorithms include monitoring human activity to study the dynamics of human contacts, such as hospitals, museums and conferences [7,34,59].

In the rest of the paper, we provide information describing the data collection and preprocessing for the purposes of network detection analysis in Sect. 3, the results of the analysis using the six algorithms described above in Sect. 4, and various insights regarding the nature of the communities detected in Sect. 5.

2   Background and Related Work

The analysis of community structure [47], is a prominent research topic in data mining and social network analysis. In the context of this paper, we focus on user interaction formalized in so-called social interaction networks [2,43,44]: These refer to user-related social networks in social media that are capturing social relations inherent in social interactions, social activities and other social phenomena which act as proxies for social user-relatedness. Essentially, social interaction networks focus on *interaction* relations between *people*, see [62, p. 37 ff.], that are the corresponding actors.

Here, we consider social media as a broader term, also including (offline) interactions between people, captured by wearable sensors. Below we summarize background and related work on methods for capturing social interactions, as well as on their community analysis.

2.1   Capturing Human Interaction Networks

As outlined above, the SocioPatterns Collaboration (See Footnote 1) developed proximity tags based on Radio Frequency Identification technology (RFID-chips). Therefore, such social interaction networks which capture offline interactions between people can be applied and analyzed in various contexts. Eagle and Pentland [27], for example, presented an analysis using proximity information collected by bluetooth devices as a proxy for human proximity. However, given the range of interaction of bluetooth devices, the detected proximity does not necessarily correspond to face-to-face contacts [10,17,31,58]. Other approaches for observing human face-to-face communication are given by the Sociometric[2]

---

[2] http://hd.media.mit.edu/badges.

and Rhythm[3] Badges [40,41,48]. However, while these record more details of the interaction, they require significantly larger devices.

Recently, the SocioPattern sensors have been used in several ubiquitous and social environments, varying from scientific conferences [22,42], hospitals [35], museums [34], schools [23,30,59], student freshman weeks [15], and scientific collaborative events [9,39]. First results of community detection in such face-to-face contact networks have been reported in [7,37,38], for example, however neglecting time and other feature-oriented analysis.

2.2   Community Detection

For the automatic detection of communities, several algorithms can be used, e.g., [28,47], for which standard methods are implemented in the igraph software package [25]. The general problem definition of community detection can be formulated as follows: given a network $G(V, E)$, find the optimal communities in closely connected groups of nodes and a moderate number of disparate outliers [61]. We applied the following algorithms from recently proposed taxonomies of numerous community detection methods [24,28]: Edge Betweenness, Info Map, Spinglass, Louvain, Label Propagation, and Leading Eigenvector. The differences between these algorithms are quite substantial:

- Edge Betweenness is a hierarchical process decomposing a given graph [47] by removing edges in decreasing order of their edge betweenness scores, i. e., identifying edges on multiple shortest paths which in many cases are bridges linking groups together.
- Infomap is based on random walks and information theoretic principles [53]. The algorithm tries to group nodes based on the shortest description length for a random walk on the graph. The description length is measured by the number of bits per vertex required to encode the path of the random walk. The best communities are those that transmit a large deal of information while requiring minimal bandwidth.
- The Spinglass algorithm originates from physics and is based on the Potts model [52]. Each node can be in a certain state and the interactions between the nodes (the edges) specify which state the node has. This process is simulated several times and nodes with the same state are seen as a community. Spinglass provides a parameter determining the cluster sizes and, thus the number of communities.
- The Louvain method is a top-down process based on modularity optimization [19]. First, small communities are identified by their modularity score [32, 47] (each node is its own community); subsequently, smaller communities are grouped together if and only if it increases the modularity score. This process is repeated until the merging of communities will no longer lead to an improved modularity score.

---

[3] http://www.rhythm.mit.edu/.

– Finally, Leading Eigenvector is modularity based and uses optimization, inspired by a technique called graph partitioning [46]. The algorithm tries to find the eigenvector that corresponds to the most positive eigenvalue of the modularity matrix. Afterwards, it divides the network into communities in harmony with the elements of the vector.

A comparison of the performance for the algorithms listed above in the context of our network analysis task is provided in Sect. 4.

3 Data Set

The data was collected during two student events organized at a university located in the south of the Netherlands. Attendees were invited to participate voluntarily. They were informed about the general aim of the study (testing the use of wearables in social situations). They all signed a consent form. The data collection was anonymized, including solely information about participant gender. For the two events, the number of male and female participants was comparable (Event 1: 12 male and 11 female; Event 2: 10 male and 9 female). The age interval was between 18 to 24 years old.

For the data collection utilizing the UBICON framework [5], we applied the following procedure: Each participant was asked to wear a SocioPatterns proximity tag. Social contact was established when the contact between two proximity tags was at least 20 s [22]. Interactions that took place within an interval of 20 s were merged if the actors remained the same, according to the commonly used threshold [22]. This resulted in a data set where accidental interactions were filtered out (e. g., two people passing each other for a total of 4 s or less). In addition, a minimum RSSI value of −85 was used as a threshold to filter out weak interactions.

Importantly, for the goals of this study, during Event 1, the male participants did not know the female participants (within a given gender cluster, the participants knew each other before the start of the event). During Event 2, all participants knew each other. The male participants of Event 1 and 2 were identical; the female participants differed between the two events.

4 Results

Below, we first analyze structural properties of the networks for Events 1/2 in Sect. 4.1. After that, we analyze the respective evolution of contacts in Sect. 4.2 stratified on time. Finally, we perform a stratification-oriented analysis – regarding the effect of gender and potential signs of homophily in Sect. 4.3.

4.1 Structural Network Analysis

Below, we first offer descriptive information regarding the differences between Event 1 and 2. As shown in Table 1, there were 25 proximity tags active during Event 1, compared to 19 during Event 2. Comparing the networks for both

Table 1. High level network statistics: Number of nodes (N) and edges (E), Average contact duration (ACD), Longest contact duration (LCD), Network diameter (Diameter), Graph density (Density), Transitivity and Average path length (APL).

Network	N	E	ACD	LCD	Diameter	Density	Transitivity	APL
Event 1	25	606	309.60	5807	5	2.02	0.47	2
Event 2	19	1239	535.70	5126	2	7.25	.60	1.51

Table 2. Average node centrality measures for both events.

Network	∅Eigenvector	∅Degree	∅Closeness	∅Betweenness
Event 1	0.21	48.48	0.02	12
Event 2	0.29	130.42	0.04	4.58

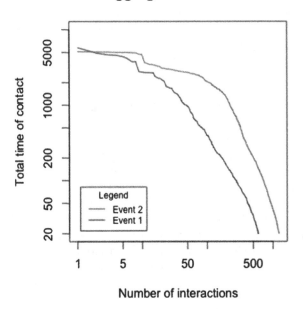

Fig. 1. Cumulated contact length (in seconds) distribution of all face-to-face contacts at Event 1 and Event 2.

events, some structural differences can be observed with respect to contact duration and the number of interactions (see Table 1).

In line with the expectation, the number of edges and the average contact length of the Event 1 network are lower than the number of edges and average contact length for Event 2 (recall that during Event 2, all participants already knew each other). Next to that, the network of Event 1 shows a diameter of 2, which indicates a small-world effect [21]. Table 2 shows several centrality measures, e. g., [20,29]: the individual measures show which participants played an important role during the events.

Interestingly, Event 2 showed a higher eigenvector centrality than Event 1. A high eigenvector score suggests that all nodes are connected to other nodes that have a high eigenvector score, which indicates that the network of Event 2 was densely inter-connected. The degree centrality of Event 2 was also higher than for Event 1, indicating that on average, the participants during Event 2 had more connections (proportionally, since the networks differ in the number of participants). Both events show a low closeness score of 0.02–0.04 indicating that each participant could be reached in relatively few steps. Presumably, this is due to the relatively small number of participants.

In Fig. 1, the aggregated contact duration of both events is shown. In line with a commonly seen power law [17,22,42], the shorter a contact was, the more likely it was to occur.

4.2   Time: Evolution of Contacts

In order to analyze the evolution of contacts, the events were divided into three intervals. In Table 3, the evolution of several measures of the network of Event 1 and 2 is shown. During each interval of the event, almost all nodes were active (see Table 3). However, differences in distribution are visible when examining the number of edges that are present in each interval. The number of active edges decreases in Event 1 in interval 2, but recovers during interval 3.

Table 3. Graph evolution metrics of Event 1 and Event 2: Number of nodes (N), edges (E), Diameter, Average path length (APL), Longest contact duration (LCD, in seconds), Average contact duration (ACD, in seconds), Density, Average Degree, Closeness and Betweenness centrality.

Event1	Periods			Event2	Periods		
	1	2	3		1	2	3
N	23	18	22	N	15	17	19
E	268	145	193	E	150	411	678
Diameter	5	4	5	Diameter	3	5	3
APL	2.41	2.24	2.50	APL	1.94	2.01	1.64
MCD	5807	2783	1009	MCD	5126	3445	1630
ACD	422.97	369.10	107.47	ACD	1041.07	916.58	193.02
Density	1.06	0.95	0.84	Density	1.43	3.02	3.95
∅Degree	23.30	16.11	17.55	∅Degree	20	48.35	71.37
∅Closeness	0.02	0.03	0.02	∅Closeness	0.04	0.03	0.03
∅Between	15.5	10.50	15.77	∅Betweenness	6.6	8.06	5.79

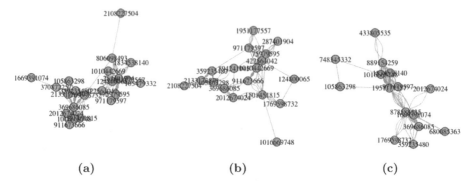

Fig. 2. Event 1 divided into three equal intervals. From left to right: (a) interval 1, (b) interval 2 and (c) interval 3.

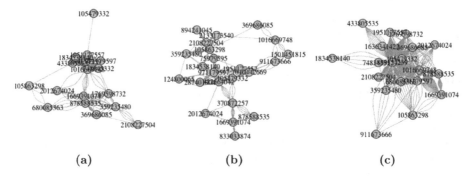

Fig. 3. Event 2 divided into three equal intervals. From left to right: (a) interval 1, (b) interval 2 and (c) interval 3.

The diameter scores confirm this observation. The diameter was lowest during interval 2 of Event 1 (indicating a small world effect) and highest during interval 1 and 3. A similar pattern to the evolution of edges and diameter is visible in the edge density and average degree.

For Event 2, the average contact duration decreases between interval 1 to interval 2 and rises again in interval 3. Similarly to Event 1, not all nodes were active during each period of the event and the number of edges increases in the second and even the third interval. Finally, in Figs. 2 and 3, the changes in the Betweenness centrality are clearly represented.

It is important to note that the results of interval 3 in Event 2 are negatively influenced by the fact that the graph becomes sparsely connected by two bridges. Especially the closeness centrality suffers from a disconnected graph since the metric calculates the distance between nodes. If the network is disconnected, the distance is infinite [49,62]. One might not notice this effect in Table 3, but the closeness centrality decline from 0.02 to 0.04 is large, since the values of the closeness centrality metric tend to span a moderately small dynamic range [49].

4.3   Gender: Analyis of Community Emergence

In order to analyze the effect of sociodemographic variables on community emergence, we focused on the effect of gender and investigated potential signs of homophily in the community formation structure. The rationale for exploring gender effects is that gender plays an important role in the social life of university undergraduates.

In Fig. 4, the networks of both events are depicted and exemplary communities are shown as detected using the Louvain algorithm. The visualization suggests that homophily likely played a role in the creation of communities during both events. The intra-group networks appear to have a higher density than the inter-group networks. We explored the effect of gender further below by means of quantitative analysis with weighting, comparing Event 1 (where male participants did not previously know the female participants) and Event 2 (where all participants have previously met).

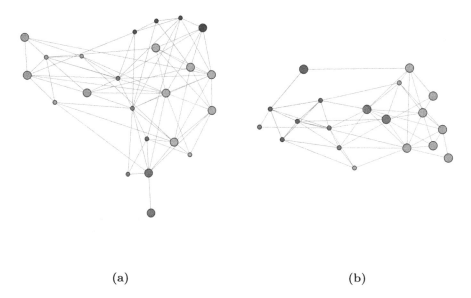

(a)   (b)

Fig. 4. Communities detected using Louvain (Modularity) where male nodes are sized big and female nodes are sized small: (a) Event 1, (b) Event 2.

First of all, in order to analyze the robustness of the community detection algorithms, we used the modularity score of the total graph and the average computational time (each algorithm was run a total of 20 times and the time was averaged). On average, the modularity score, which gives insight into the strength of the divisions of a network into communities, was 0.27 ($SD = 0.19$) for Event 1 and 0.08 ($SD = 0.07$) for Event 2. No algorithm managed to determine communities with a modularity higher than 0.46 (see Table 4). The Spinglass

Table 4. Comparison of the applied community detection algorithms for both events. The number of communities, modularity scores, and the computational time are averaged over 20 simulations.

Algorithm	Parameter	Event 1	Event 2
Edge Betweenness	No. of Communities	6	8
	Modularity	0.24	0.03
	Computational Time	3.03	12.07
Info Map	No. of Communities	10	7
	Modularity	0.04	0.09
	Computational Time	0.37	0.41
Label Prop	No. of Communities	4	1
	Modularity	0.46	0.00
	Computational Time	0.05	0.06
Leading Eigenvector	No. of Communities	5	3
	Modularity	0.39	0.15
	Computational Time	1.44	1.16
Louvain	No. of Communities	4	3
	Modularity	0.46	0.17
	Computational Time	0.07	0.07
Spinglass	No. of Communities	4	3
	Modularity	0.07	0.03
	Computational Time	11.54	10.93

Table 5. Comparison of community structures using NMI on event 1.

Comparisons								
Edge Betweenness	Info Map	0.58	Info Map	Edge Betweenness	0.58	Spinglass	Edge Betweenness	0.86
	Spinglas	0.86		Spinglas	0.53		Info Map	0.53
	Louvain	0.92		Louvain	0.56		Louvain	0.83
	Label Prop	0.79		Label Prop	0.51		Label Prop	0.68
	Leading Eig	0.80		Leading Eig	0.61		Leading Eig	0.73
Louvain	Edge Betweenness	0.92	Label Prop	Leading Eig	0.72	Leading Eig	Edge Betweenness	0.80
	Info Map	0.56		Edge Betweenness	0.79		Info Map	0.61
	Spinglas	0.83		Info Map	0.51		Spinglass	0.73
	Label Prop	0.87		Spinglass	0.68		Louvain	0.86
	Leading Eig	0.86		Louvain	0.87		Label Prop	0.72

algorithm performed worst on both events and took the longest to compute. The Louvain and Label Propagation algorithms showed the highest modularity scores for Event 1 and achieved this result in a short computational time.

Table 6. Comparison of community structures using NMI on event 2.

Comparisons								
Edge Betweenness	Info Map	0.53	Info Map	Edge Betweenness	0.53	Spinglass	Edge Betweenness	0.55
	Spinglas	0.55		Spinglas	0.37		Info Map	0.37
	Louvain	0.51		Louvain	0.24		Louvain	0.20
	Label Prop	0.00		Label Prop	0.00		Label Prop	0.00
	Leading Eig	0.56		Leading Eig	0.33		Leading Eig	0.35
Louvain	Edge Betweenness	0.52	Label Prop	Leading Eig	0.00	Leading Eig	Edge Betweenness	0.56
	Info Map	0.24		Edge Betweenness	0.00		Info Map	0.33
	Spinglas	0.20		Info Map	0.00		Spinglass	0.35
	Label Prop	0.00		Spinglass	0.00		Louvain	0.51
	Leading Eig	0.51		Louvain	0.00		Label Prop	0.00

We further examined the contribution of individual algorithms using the normalized mutual information metric (NMI) [26]. The more basic variant mutual information (MI) is a measure of the mutual dependence between two variables [26]. Since NMI is normalized, we can measure and compare the NMI score of the resulted networks in regard to their calculated communities (NMI = 0 means no mutual information, NMI = 1 means perfect correlation). Tables 5 and 6 show the NMI scores of each algorithm for Event 1 and 2.

Second, using a method inspired by [16] for the analysis of social interaction networks, we examined stratification according to gender. However, compared to [16] we do not focus on the global contact structure – in contrast we assessed

Table 7. Effect of inter-group weighting on community detection results: Community detection of event 1 and 2 with weight 2 where groups is the number of communities, mod the Modularity and NMI the normalized mutual information.

Method	Network 1		Network 2		
	Groups	Modularity	Groups	Modularity	NMI
Infomap unweighted	11	0.12	8	−0.10	0.72
Infomap	10	0.06	8	−0.04	0.67
Label Propagation unweighted	4	0.48	1	0.00	0.77
Label Propagation	4	0.43	1	0.00	1.00
Leading Eigenvector unweighted	5	0.39	3	0.16	0.83
Leading Eigenvector	5	0.39	2	0.17	0.65
Louvain unweighted	4	0.48	3	0.17	1.00
Louvain	4	0.45	2	0.17	0.21
Spinglass unweighted	4	0.07	3	0.03	0.57
Spinglass	4	0.05	2	0.03	0.48

Table 8. Effect of inter-group weighting on community detection results: Community detection of event 1 and 2 with weight 3 where groups is the number of communities, mod the Modularity and NMI the normalized mutual information.

Method	Network 1		Network 2		
	Groups	Modularity	Groups	Modularity	NMI
Infomap unweighted	8	0.05	7	−0.06	0.67
Infomap	10	0.10	7	−0.07	0.71
Label Propagation unweighted	4	0.48	1	0.00	0.65
Label Propagation	6	0.41	1	0.00	1.00
Leading Eigenvector unweighted	5	0.39	3	0.16	0.83
Leading Eigenvector	5	0.40	2	0.18	0.65
Louvain unweighted	4	0.48	3	0.17	0.77
Louvain	4	0.43	2	0.18	0.21
Spinglass unweighted	4	0.07	3	0.03	0.56
Spinglass	4	0.04	3	0.02	0.42

the individual communities. That is, we compared the contribution of gender to community creation during Event 1 by assigning additional weight to male-female contacts, i. e., in a sense penalizing contacts between participants who were already acquainted.

We summarize the results of applying the different community detection algorithms in Table 7, where we selected those algorithms handling edge weights. That is, as outlined above, mixed-gender edges linking participants with different genders received a larger weight. These results show that while the size of the detected communities is largely unimpacted by the weighting, overall, there is a slight decrease in the modularity score after the weights have been assigned (for Event 1). For Event 2, on the contrary, penalizing intra-group contacts leads to a lower number of communities detected by the selected algorithms, which can also be seen in effect when considering the lowered modularity values. This result confirms the observation based on the visual analysis of Fig. 4: Here, particularly in the context where male and female participants were mutually acquainted, they tended to establish contacts with members of the same gender rather than seeking contact with members of the opposite gender. This is then also reflected in the community structure.

Lastly, we increased the weight from 2 to 3 in order to investigate the effect of this parameter even more. When considering the higher threshold, we observe the same trends as for the lower threshold. However, no significant changes in the number of communities detected and their modularity seem to arise, as can be observed in Table 8.

5   Conclusions

In the current study, we explored the use of RFID chips with proximity detection as possible means to collect face-to-face contact data during two events. Here, we focussed on the evolution of contacts and the dynamics of the community structure. The two events were likely to give rise to different interaction patterns since during Event 1, male and female participants were not previously acquainted with one another, contrary to Event 2. We performed an analysis of the contact graphs and examined the evolution of contacts by dividing the contact networks into three different intervals.

Furthermore, in order to automatically detect communities, six conceptually different community detection algorithms were used. To analyze the quality of the communities that were detected, their modularity score, the number of detected communities, the computational time each algorithm needed and normalized mutual information were examined.

For the analysis on the evolution of the face-to-face contacts and the respective interactions, we observed that during both events, face-to-face contacts tended to develop in a parabolic manner. This parabolic tendency is contradictory to the linear patterns other studies have found [37, 38]. The most likely explanation is that the linear patterns were found for events with a highly structured program (for example, the end of the presentation of the key speaker would likely result in a peak of interactions at the end of the event).

With respect to community detection, the Label Propagation and Louvain algorithms showed the most promising results in the network of Event 1. Both their modularity scores ranked the highest while demanding the least computational time. The lack of results for the Leading Eigenvector and Label Propagation algorithms is in line with earlier studies [50]. It is interesting to note that the algorithm performs poorly in both an experimental (in most studies the networks are randomly generated) and the real-life setting.

Finally, we observed effects of homophily, particularly during the event where participants of both genders knew each other from past interactions (Event 2). After splitting up the events into gender-stratified networks, the female ones presented significantly higher modularity scores. To address limitations given by the size of the set of participants, we aim to perform a larger follow-up study. Here, we intend to increase the size of the networks to make the results further generalizable and also to include ground truth information by means of video information and participant recollection.

Other interesting directions for future work are given by including spatial/localization information for analyzing spatio-temporal patterns, e. g., [11, 57] as well as collecting more descriptive information in order to enable community detection on attributed networks, e. g., [8]. This enables the detection of characteristic and interpretable profiles, e. g., [3, 8, 12–14, 51, 60], based on descriptive local community patterns and the analysis of feature-rich networks [4, 33, 36].

References

1. Atzmueller, M.: Mining social media: key players, sentiments, and communities. WIREs Data Min. Knowl. Discov. **2**(5), 411–419 (2012)
2. Atzmueller, M.: Data mining on social interaction networks. J. Data Min. Digit. Hum. **1** (2014)
3. Atzmueller, M.: Compositional subgroup discovery on attributed social interaction networks. In: Soldatova, L., Vanschoren, J., Papadopoulos, G., Ceci, M. (eds.) DS 2018. LNCS (LNAI), vol. 11198, pp. 259–275. Springer, Cham (2018). https://doi.org/10.1007/978-3-030-01771-2_17
4. Atzmueller, M.: Perspectives on model-based anomalous link pattern mining on feature-rich social interaction networks. In: Proceedings of WWW 2019 (Companion). IW3C2/ACM (2019)
5. Atzmueller, M., et al.: Ubicon and its applications for ubiquitous social computing. New Rev. Hypermedia Multimed. **20**(1), 53–77 (2014)
6. Atzmueller, M., et al.: Enhancing social interactions at conferences. IT - Inf. Technol. **53**(3), 101–107 (2011)
7. Atzmueller, M., Doerfel, S., Hotho, A., Mitzlaff, F., Stumme, G.: Face-to-face contacts at a conference: dynamics of communities and roles. In: Atzmueller, M., Chin, A., Helic, D., Hotho, A. (eds.) MSM/MUSE 2011. LNCS (LNAI), vol. 7472, pp. 21–39. Springer, Heidelberg (2012). https://doi.org/10.1007/978-3-642-33684-3_2
8. Atzmueller, M., Doerfel, S., Mitzlaff, F.: Description-oriented community detection using exhaustive subgroup discovery. Inf. Sci. **329**, 965–984 (2016)
9. Atzmueller, M., Ernst, A., Krebs, F., Scholz, C., Stumme, G.: On the evolution of social groups during coffee breaks. In: Proceedings of WWW 2014 (Companion), pp. 631–636. IW3C2/ACM (2014)
10. Atzmueller, M., Hilgenberg, K.: Towards capturing social interactions with SDCF: an extensible framework for mobile sensing and ubiquitous data collection. In: Proceedings of the 4th International Workshop on Modeling Social Media (MSM 2013), Hypertext 2013. ACM Press, New York (2013)
11. Atzmueller, M., Lemmerich, F.: Exploratory pattern mining on social media using geo-references and social tagging information. IJWS **2**(1/2), 80–112 (2013)
12. Atzmueller, M., Lemmerich, F., Krause, B., Hotho, A.: Who are the spammers? Understandable local patterns for concept description. In: Proceedings of the 7th Conference on Computer Methods and Systems, Krakow, Poland (2009)
13. Atzmueller, M., Mitzlaff, F.: Efficient descriptive community mining. In: Proceedings of the 24th International FLAIRS Conference, pp. 459–464. AAAI Press, Palo Alto (2011)
14. Atzmueller, M., Puppe, F., Buscher, H.P.: Profiling examiners using intelligent subgroup mining. In: Proceedings of the 10th International Workshop on Intelligent Data Analysis in Medicine and Pharmacology, Aberdeen, Scotland, pp. 46–51 (2005)
15. Atzmueller, M., Thiele, L., Stumme, G., Kauffeld, S.: Analyzing group interaction on networks of face-to-face proximity using wearable sensors. In: Proceedings of the IEEE International Conference on Future IoT Technologies, Boston, MA, USA. IEEE (2018)
16. Atzmueller, M., Thiele, L., Stumme, G., Kauffeld, S.: Analyzing group interaction on networks of face-to-face proximity using wearable sensors. In: IEEE International Conference on Future IoT Technologies. IEEE (2018)

17. Barrat, A., Cattuto, C., Colizza, V., Pinton, J.F., Broeck, W.V.d., Vespignani, A.: High resolution dynamical mapping of social interactions with active RFID. arXiv preprint: arXiv:0811.4170 (2008)
18. Bloemheuvel, S., Atzmueller, M., Postma, M.: Evolution of contacts and communities in social interaction networks of face-to-face proximity. In: Proceedings of BNAIC. Jheronimus Academy of Data Science, Den Bosch, The Netherlands (2018)
19. Blondel, V.D., Guillaume, J.L., Lambiotte, R., Lefebvre, E.: Fast unfolding of communities in large networks. J. Stat. Mech. Theory Exp. **2008**(10), P10008 (2008)
20. Bonacich, P.: Technique for analyzing overlapping memberships. Sociol. Methodol. **4**, 176–185 (1972)
21. Brust, M.R., Rothkugel, S.: Small worlds: strong clustering in wireless networks. arXiv preprint: arXiv:0706.1063 (2007)
22. Cattuto, C., Van den Broeck, W., Barrat, A., Colizza, V., Pinton, J.F., Vespignani, A.: Dynamics of person-to-person interactions from distributed RFID sensor networks. PloS ONE **5**(7), e11596 (2010)
23. Ciavarella, C., Fumanelli, L., Merler, S., Cattuto, C., Ajelli, M.: School closure policies at municipality level for mitigating influenza spread: a model-based evaluation. BMC Infect. Dis. **16**(1), 576 (2016)
24. Coscia, M., Giannotti, F., Pedreschi, D.: A classification for community discovery methods in complex networks. Stat. Anal. Data Min. ASA Data Sci. J. **4**(5), 512–546 (2011)
25. Csardi, G., Nepusz, T.: The igraph software package for complex network research. Int. J. Complex Syst. **1695**(5), 1–6 (2006). http://igraph.org
26. Danon, L., Diaz-Guilera, A., Duch, J., Arenas, A.: Comparing community structure identification. J. Stat. Mech. Theory Exp. **2005**(09), P09008 (2005)
27. Eagle, N., Pentland, A.S., Lazer, D.: Inferring friendship network structure by using mobile phone data. PNAS **106**(36), 15274–15278 (2009)
28. Fortunato, S.: Community detection in graphs. Phys. Rep. **486**(3–5), 75–174 (2010)
29. Freeman, L.C.: Centrality in social networks conceptual clarification. Soc. Netw. **1**(3), 215–239 (1978)
30. Gemmetto, V., Barrat, A., Cattuto, C.: Mitigation of infectious disease at school: targeted class closure vs school closure. BMC Infect. Dis. **14**(1), 695 (2014). https://doi.org/10.1186/PREACCEPT-6851518521414365
31. Genois, M., Barrat, A.: Can co-location be used as a proxy for face-to-face contacts? EPJ Data Sci. **7**(1), 11 (2018)
32. Girvan, M., Newman, M.E.: Community structure in social and biological networks. Proc. Natl. Acad. Sci. **99**(12), 7821–7826 (2002)
33. Interdonato, R., Atzmueller, M., Gaito, S., Kanawati, R., Largeron, C., Sala, A.: Feature-rich networks: going beyond complex network topologies. Appl. Netw. Sci. **4**(4) (2019)
34. Isella, L., et al.: Close encounters in a pediatric ward: measuring face-to-face proximity and mixing patterns with wearable sensors. PloS ONE **6**(2), e17144 (2011)
35. Isella, L., Stehlé, J., Barrat, A., Cattuto, C., Pinton, J.F., Van den Broeck, W.: What's in a crowd? Analysis of face-to-face behavioral networks. J. Theor. Biol. **271**(1), 166–180 (2011)
36. Kanawati, R., Atzmueller, M.: Modeling and mining feature-rich networks. In: Proceedings of WWW 2019 (Companion). IW3C2/ACM (2019)
37. Kibanov, M., Atzmueller, M., Scholz, C., Stumme, G.: On the evolution of contacts and communities in networks of face-to-face proximity. In: Proceedings of IEEE CPSCom, pp. 993–1000. IEEE (2013)

38. Kibanov, M., Atzmueller, M., Scholz, C., Stumme, G.: Temporal evolution of contacts and communities in networks of face-to-face human interactions. Sci. China Inf. Sci. **57**(3), 1–17 (2014)
39. Kibanov, M., Heiberger, R., Roedder, S., Atzmueller, M., Stumme, G.: Social studies of scholarly live with sensor-based ethnographic observations. Scientometrics **119**(3), 1387–1428 (2019)
40. Kim, T., McFee, E., Olguin, D.O., Waber, B., Pentland, A.: Sociometric badges: using sensor technology to capture new forms of collaboration. J. Organ. Behav. **33**(3), 412–427 (2012)
41. Lederman, O., Mohan, A., Calacci, D., Pentland, A.S.: Rhythm: a unified measurement platform for human organizations. IEEE MultiMedia **25**(1), 26–38 (2018)
42. Macek, B.E., Scholz, C., Atzmueller, M., Stumme, G.: Anatomy of a conference. In: Proceedings of ACM Hypertext, pp. 245–254. ACM (2012)
43. Mitzlaff, F., Atzmueller, M., Hotho, A., Stumme, G.: The social distributional hypothesis: a pragmatic proxy for homophily in online social networks. Soc. Netw. Anal. Min. **4**(1), 216 (2014)
44. Mitzlaff, F., Atzmueller, M., Stumme, G., Hotho, A.: Semantics of User Interaction in Social Media. In: Ghoshal, G., Poncela-Casasnovas, J., Tolksdorf, R. (eds.) Complex Networks IV. SCI, vol. 476, pp. 13–25. Springer, Berlin (2013). https://doi.org/10.1007/978-3-642-36844-8_2
45. Mossong, J., et al.: Social contacts and mixing patterns relevant to the spread of infectious diseases. PLoS Med. **5**(3), e74 (2008)
46. Newman, M.E.: Finding community structure in networks using the eigenvectors of matrices. Phys. Rev. E **74**(3), 036104 (2006)
47. Newman, M.E., Girvan, M.: Finding and evaluating community structure in networks. Phys. Rev. E **69**(2), 026113 (2004)
48. Olguin, D.O., Pentland, A.S.: Sociometric badges: state of the art and future applications. In: Doctoral Colloquium Presented at IEEE 11th International Symposium on Wearable Computers, Boston, MA (2007)
49. Opsahl, T., Agneessens, F., Skvoretz, J.: Node centrality in weighted networks: generalizing degree and shortest paths. Soc. Netw. **32**(3), 245–251 (2010)
50. Orman, G.K., Labatut, V., Cherifi, H.: On accuracy of community structure discovery algorithms. arXiv preprint: arXiv:1112.4134 (2011)
51. Pool, S., Bonchi, F., van Leeuwen, M.: Description-driven community detection. Trans. Intell. Syst. Technol. **5**(2), 1–28 (2014)
52. Reichardt, J., Bornholdt, S.: Statistical mechanics of community detection. Phys. Rev. E **74**(1), 016110 (2006)
53. Rosvall, M., Bergstrom, C.T.: Maps of random walks on complex networks reveal community structure. PNAS **105**(4), 1118–1123 (2008)
54. Scholz, C., Atzmueller, M., Barrat, A., Cattuto, C., Stumme, G.: New insights and methods for predicting face-to-face contacts. In: Proceedings of the 7th International AAAI Conference on Weblogs and Social Media. AAAI Press, Palo Alto (2013)
55. Scholz, C., Atzmueller, M., Kibanov, M., Stumme, G.: Predictability of evolving contacts and triadic closure in human face-to-face proximity networks. Soc. Netw. Anal. Min. **4**(1), 217 (2014)
56. Scholz, C., Atzmueller, M., Stumme, G.: On the predictability of human contacts: influence factors and the strength of stronger ties. In: Proceedings of IEEE SocialCom, pp. 312–321. IEEE (2012)

57. Scholz, C., Doerfel, S., Atzmueller, M., Hotho, A., Stumme, G.: Resource-aware on-line RFID localization using proximity data. In: Gunopulos, D., Hofmann, T., Malerba, D., Vazirgiannis, M. (eds.) ECML PKDD 2011, Part III. LNCS (LNAI), vol. 6913, pp. 129–144. Springer, Heidelberg (2011). https://doi.org/10.1007/978-3-642-23808-6_9

58. Starnini, M., Lepri, B., Baronchelli, A., Barrat, A., Cattuto, C., Pastor-Satorras, R.: Robust modeling of human contact networks across different scales and proximity-sensing techniques. In: Ciampaglia, G.L., Mashhadi, A., Yasseri, T. (eds.) SocInfo 2017, Part I. LNCS, vol. 10539, pp. 536–551. Springer, Cham (2017). https://doi.org/10.1007/978-3-319-67217-5_32

59. Stehlé, J., et al.: High-resolution measurements of face-to-face contact patterns in a primary school. PloS ONE **6**(8), e23176 (2011)

60. Tumminello, M., Micciche, S., Lillo, F., Varho, J., Piilo, J., Mantegna, R.N.: Community characterization of heterogeneous complex systems. J. Stat. Mech. Theory Exp. **2011**(01), P01019 (2011)

61. Wang, M., Wang, C., Yu, J.X., Zhang, J.: Community detection in social networks: an in-depth benchmarking study with a procedure-oriented framework. Proc. VLDB Endow. **8**(10), 998–1009 (2015)

62. Wasserman, S., Faust, K.: Social Network Analysis: Methods and Applications, vol. 8. Cambridge University Press, Cambridge (1994)

63. Zhang, Y., Wang, L., Zhang, Y.Q., Li, X.: Towards a temporal network analysis of interactive WiFi users. EPL (Europhys. Lett.) **98**(6), 68002 (2012)

Analyzing Big Data Streams with Apache SAMOA

Nicolas Kourtellis[1], Gianmarco de Francisci Morales[2], and Albert Bifet[3,4(✉)]

[1] Telefonica Research, Barcelona, Spain
nicolas.kourtellis@telefonica.com
[2] ISI Foundation, Turin, Italy
gdfm@acm.org
[3] LTCI, Télécom Paris, IP-Paris, Paris, France
albert.bifet@telecom-paristech.fr
[4] University of Waikato, Hamilton, New Zealand

Abstract. Apache Apache SAMOA (SCALABLE ADVANCED MASSIVE ONLINE ANALYSIS) is an open-source platform for mining big data streams. Big data is defined as datasets whose size is beyond the ability of typical software tools to capture, store, manage and analyze, due to the time and memory complexity. Velocity is one of the main properties of big data. Apache Apache SAMOA provides a collection of distributed streaming algorithms for the most common data mining and machine learning tasks such as classification, clustering, and regression, as well as programming abstractions to develop new algorithms. It features a pluggable architecture that allows it to run on several distributed stream processing engines such as Apache Flink, Apache Storm, Apache Samza, and Apache Apex. Apache Apache SAMOA is written in Java and is available at https://samoa.incubator.apache.org/ under the Apache Software License version 2.0.

1 Introduction

Big data are "data whose characteristics force us to look beyond the traditional methods that are prevalent at the time" [15]. Social media are one of the largest and most dynamic sources of data. These data are not only very large due to their fine grain, but also being produced continuously. Furthermore, such data are nowadays produced by users in different social environments and via a multitude of devices. For these reasons, data from social media and ubiquitous social environments are perfect examples of the challenges posed by big data.

Currently, there are two main ways to deal with these challenges: streaming algorithms and distributed computing (e.g., MapReduce). Apache SAMOA aims at satisfying the future needs for big data stream mining by combining the two approaches in a single platform under an open source umbrella [8].

Data mining and machine learning are well established techniques among social media companies and startups for mining ubiquitous and social environments. Online content analysis for detecting aggression [5], stock trade volume

© Springer Nature Switzerland AG 2019
M. Atzmueller et al. (Eds.): MUSE 2015/MSM 2015/MSM 2016, LNAI 11406, pp. 44–67, 2019.
https://doi.org/10.1007/978-3-030-34407-8_3

prediction [4], online spam detection [6], recommendation [10] and personalization are just a few of the applications made possible by mining the huge quantity of data available nowadays. Just think of Facebook's relevance algorithm for the news feed for a famous example.

The usual pipeline for mining and modeling data (what "data scientists" do) involves taking a sample from production data, cleaning and preprocessing it to make it amenable to modeling, training a model for the task at hand, and finally deploying it to production. The final output of this process is a pipeline that needs to run (and be maintained) periodically to keep the model up to date.

In order to cope with web-scale datasets, data scientists have resorted to *parallel and distributed computing*. MapReduce [9] is currently the de-facto standard programming paradigm in this area, mostly thanks to the popularity of Hadoop[1], an open source implementation of MapReduce started at Yahoo. Hadoop and its ecosystem (e.g., Mahout[2]) have proven to be an extremely successful platform to support the aforementioned process at web scale.

However, nowadays most data is generated in the form of a stream, especially when dealing with social media. Batch data is just a snapshot of streaming data obtained in an interval of time. Researchers have conceptualized and abstracted this setting in the *streaming model*. In this model data arrive at high speed, one instance at a time, and algorithms must process it in one pass under very strict constraints of space and time. Extracting knowledge from these massive data streams to perform dynamic network analysis [16] or to create predictive models [1], and using them, e.g., to choose a suitable business strategy, or to improve healthcare services, can generate substantial competitive advantages. Many applications need to process incoming data and react on-the-fly by using comprehensible prediction mechanisms (e.g., card fraud detection) and, thus, streaming algorithms make use of probabilistic data structures to give fast and approximate answers.

On the one hand, MapReduce is not suitable to express streaming algorithms. On the other hand, traditional sequential online algorithms are limited by the memory and bandwidth of a single machine. *Distributed stream processing engines* (DSPEs) are a new emergent family of MapReduce-inspired technologies that address this issue. These engines allow to express parallel computation on streams, and combine the scalability of distributed processing with the efficiency of streaming algorithms. Examples include Storm[3], Flink[4], and Samza[5].

Alas, currently there is no common solution for mining big data streams, that is, for running data mining and machine learning algorithms on a distributed stream processing engine. The goal of Apache SAMOA is to fill this gap, as exemplified by Fig. 1(left).

---

[1] http://hadoop.apache.org.
[2] http://mahout.apache.org.
[3] http://storm.apache.org.
[4] https://flink.apache.org/.
[5] http://samza.apache.org/.

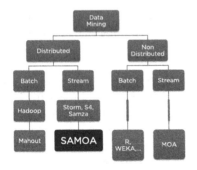 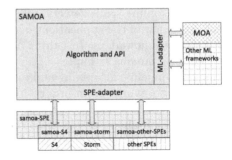

Fig. 1. (Left) Taxonomy of data mining and machine learning tools. (Right) High level architecture of Apache SAMOA.

2   Description

Apache SAMOA (SCALABLE ADVANCED MASSIVE ONLINE ANALYSIS) is a platform for mining big data streams [7]. For a simple analogy, think of Apache SAMOA as Mahout for streaming. As most of the rest of the big data ecosystem, it is written in Java.

Apache SAMOA is both a framework and a library. As a framework, it allows the algorithm developer to abstract from the underlying execution engine, and therefore reuse their code on different engines. It features a pluggable architecture that allows it to run on several distributed stream processing engines such as Storm, Flink, and Samza. This capability is achieved by designing a minimal API that captures the essence of modern DSPEs. This API also allows to easily write new bindings to port Apache SAMOA to new execution engines. Apache SAMOA takes care of hiding the differences of the underlying DSPEs in terms of API and deployment.

As a library, Apache SAMOA contains implementations of state-of-the-art algorithms for distributed machine learning on streams. For classification, Apache SAMOA provides a Vertical Hoeffding Tree (VHT), a distributed streaming version of a decision tree. For clustering, it includes an algorithm based on CluStream. For regression, HAMR, a distributed implementation of Adaptive Model Rules. The library also includes meta-algorithms such as bagging and boosting. The platform is intended to be useful for both research and real world deployments.

3   High Level Architecture

We identify three types of Apache SAMOA users:

1. Platform users, who use available ML algorithms without implementing new ones.

2. ML developers, who develop new ML algorithms on top of Apache SAMOA and want to be isolated from changes in the underlying SPEs.
3. Platform developers, who extend Apache SAMOA to integrate more DSPEs into Apache SAMOA.

There are three important design goals of Apache SAMOA:

1. **Flexibility** in terms of developing new ML algorithms or reusing existing ML algorithms from other frameworks.
2. **Extensibility** in terms of porting Apache SAMOA to new DSPEs.
3. **Scalability** in terms of handling an ever increasing amount of data.

Figure 1(right) shows the high-level architecture of Apache SAMOA which attempts to fulfil the aforementioned design goals. The *algorithm* layer contains existing distributed streaming algorithms that have been implemented in Apache SAMOA. This layer enables platform users to easily use the existing algorithm on any DSPE of their choice.

The *application programming interface* (API) layer consists of primitives and components that facilitate ML developers when implementing new algorithms. The *ML-adapter* layer allows ML developers to integrate existing algorithms in MOA or other ML frameworks into Apache SAMOA. The API layer and ML-adapter layer in Apache SAMOA fulfill the flexibility goal since they allow ML developers to rapidly develop algorithms.

Next, the *DSPE-adapter* layer supports platform developers in integrating new DSPEs into Apache SAMOA. To perform the integration, platform developers should implement the *samoa-SPE* layer as shown in Fig. 1(right). Currently, Apache SAMOA is equipped with four adapters: the *samoa-Storm* adapter for Storm, the *samoa-Flink* adapter for Flink, the *samoa-Samza* adapter for Samza and the *samoa-S4* adapter for S4 (as of now, decommissioned). To satisfy the extensibility goal, the DSPE-adapter layer decouples DSPEs and ML algorithms implementations in Apache SAMOA, so that platform developers are able to easily integrate more DSPEs.

The last goal, scalability, implies that Apache SAMOA should be able to scale to cope ever increasing amount of data. To fulfill this goal, Apache SAMOA utilizes modern DSPEs to execute its ML algorithms. The reason for using modern DSPEs such as Storm, Flink, Samza and S4 in Apache SAMOA is that they are designed to provide horizontal scalability to cope with web-scale streams.

4 System Design

An algorithm in Apache SAMOA is represented by a directed graph of nodes that communicate via messages along streams which connect pairs of nodes. Borrowing the terminology from Storm, this graph is called a *Topology*. Each node in a Topology is a *Processor* that sends messages through a *Stream*. A Processor is a container for the code implementing the algorithm. A Stream can have a single source but several destinations (akin to a pub-sub system).

A Topology is built by using a *TopologyBuilder*, which connects the various pieces of user code to the platform code and performs the necessary bookkeeping in the background. The following is a code snippet to build a topology that joins two data streams in Apache SAMOA:

```
TopologyBuilder builder = new TopologyBuilder();
Processor sourceOne = new SourceProcessor();
builder.addProcessor(sourceOne);
Stream streamOne = builder
    .createStream(sourceOne);

Processor sourceTwo = new SourceProcessor();
builder.addProcessor(sourceTwo);
Stream streamTwo = builder
    .createStream(sourceTwo);

Processor join = new JoinProcessor();
builder.addProcessor(join)
    .connectInputShuffle(streamOne)
    .connectInputKey(streamTwo);
```

A *Task* is an execution entity, similar to a job in Hadoop. A Topology is instantiated inside a Task to be run by Apache SAMOA. An example of a Task is *PrequentialEvaluation*, a classification task where each instance is used for testing first, and then for training.

A message or an event is called *Content Event* in Apache SAMOA. As the name suggests, it is an event which contains content which needs to be processed by the processors. Finally, a *Processing Item* is a hidden physical unit of the topology and is just a wrapper of Processor. It is used internally, and it is not accessible from the API.

5   Machine Learning Algorithms

In Apache SAMOA there are currently three types of algorithms performing basic machine learning functionalities such as classification via a decision tree (VHT), clustering (CluStream) and regression rules (AMR).

The Vertical Hoeffding Tree (VHT) [17] is a distributed extension of the VFDT [11]. The VHT uses vertical parallelism to split the workload across several machines. Vertical parallelism leverages the parallelism across attributes in the same example, rather than across different examples in the stream. In practice, each training example is routed through the tree model to a leaf. There, the example is split into its constituting attributes, and each attribute is sent to a different Processor instance that keeps track of sufficient statistics.

This architecture has two main advantages over one based on horizontal parallelism. First, attribute counters are not replicated across several machines, thus reducing the memory footprint. Second, the computation of the fitness of an attribute for a split decision (via, e.g., entropy or information gain) can be performed in parallel. The drawback is that in order to get good performances, there must be sufficient inherent parallelism in the data. That is, the VHT works best for sparse data.

Apache SAMOA includes a distributed version of CluStream, an algorithm for clustering evolving data streams. CluStream keeps a small summary of the data received so far by computing micro-clusters online. These micro-clusters are further refined to create macro-clusters by a micro-batch process, which is triggered periodically. The period is configured via a command line parameter (e.g., every 10 000 examples).

For regression, Apache SAMOA provides a distributed implementation of Adaptive Model Rules [19]. The algorithm, HAMR, uses a hybrid of vertical and horizontal parallelism to distribute AMRules on a cluster.

Apache SAMOA also includes adaptive implementations of ensemble methods such as bagging and boosting. These methods include state-of-the-art change detectors such as ADWIN, DDM, EDDM, and Page-Hinckley [12]. These meta-algorithms are most useful in conjunction with external single-machine classifiers which can be plugged in Apache SAMOA. For instance, open-source connectors for MOA [3] are provided separately by the Apache SAMOA-MOA package[6].

The following listing shows how to download, build and run Apache SAMOA.

```
# download and build SAMOA
git clone http://git.apache.org/incubator-samoa.git
cd incubator-samoa
mvn package

# download the Forest Cover Type dataset
wget "http://downloads.sourceforge.net/project/moa-datastream
    /Datasets/Classification/covtypeNorm.arff.zip"
unzip "covtypeNorm.arff.zip"

# run SAMOA in local mode
bin/samoa local target/SAMOA-Local-0.4.0-SNAPSHOT.jar "
    PrequentialEvaluation -l classifiers.ensemble.Bagging -s
    (ArffFileStream -f covtypeNorm.arff) -f 100000"
```

6   Vertical Hoeffding Tree

We explain the details of the *Vertical Hoeffding Tree* [17], which is a data-parallel, distributed version of the Hoeffding tree. The *Hoeffding tree* [11] (a.k.a. VFDT) is a streaming decision tree learner with statistical guarantees. In particular, by leveraging the Chernoff-Hoeffding bound [13], it guarantees that the learned model is asymptotically close to the model learned by the batch greedy heuristic, under mild assumptions. The learning algorithm is very simple. Each leaf keeps track of the statistics for the portion of the stream it is reached by, and

---

[6] https://github.com/samoa-moa/samoa-moa.

computes the best two attributes according to the splitting criterion. Let ΔG be the difference between the value of the functions that represent the splitting criterion of these two attributes. Let ϵ be a quantity that depends on a user-defined confidence parameter δ, and that decreases with the number of instances processed. When $\Delta G > \epsilon$, then the currently best attribute is selected to split the leaf. The Hoeffding bound guarantees that this choice is the correct one with probability larger than $1 - \delta$.

In this section, first, we describe the parallelization and the ideas behind our design choice. Then, we present the engineering details and optimizations we employed to obtain the best performance.

6.1 Vertical Parallelism

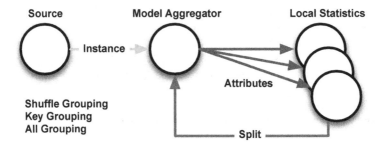

Fig. 2. High level diagram of the VHT topology.

Data parallelism is a way of distributing work across different nodes in a parallel computing environment such as a cluster. In this setting, each node executes the same operation on different parts of the dataset. Contrast this definition with task parallelism (aka pipelined parallelism), where each node executes a different operator and the whole dataset flows through each node at different stages. When applicable, data parallelism is able to scale to much larger deployments, for two reasons: (i) data has usually much higher intrinsic parallelism that can be leveraged compared to tasks, and (ii) it is easier to balance the load of a data-parallel application compared to a task-parallel one. These attributes have led to the high popularity of the currently available DSPEs. For these reasons, we employ data parallelism in the design of VHT.

In machine learning, it is common to think about data in matrix form. A typical linear classification formulation requires to find a vector x such that $A \cdot x \approx b$, where A is the data matrix and b is a class label vector. The matrix A is $n \times m$-dimensional, with n being the number of data instances and m being the number of attributes of the dataset.

There are two ways to *slice* the data matrix to obtain data parallelism: by row or column. The former is called *horizontal parallelism*, the latter *vertical parallelism*. With horizontal parallelism, data instances are independent from each other, and can be processed in isolation while considering all available attributes. With vertical parallelism, instead, attributes are considered independent from each other.

The fundamental operation of the algorithm is to accumulate statistics n_{ijk} (i.e., counters) for triplets of attribute i, value j, and class k, for each leaf l of the tree. The counters for each leaf are independent, so let us consider the case for a single leaf. These counters, together with the learned tree structure, constitute the state of the VHT algorithm.

Different kinds of parallelism distribute the counters across computing nodes in different ways. With horizontal parallelism [2], the instances are distributed randomly, thus multiple instances of the same counter can exist on several nodes. On the other hand, when using vertical parallelism, the counters for one attribute are grouped on a single node.

This latter design has several advantages. First, by having a single copy of the counter, the memory requirements for the model are the same as in the sequential version. In contrast, with horizontal parallelism a single attribute may be tracked on every node, thus the memory requirements grow linearly with the parallelism level. Second, by having each attribute tracked independently, the computation of the split criterion can be performed in parallel by several nodes. Conversely, with horizontal partitioning the algorithm needs to (centrally) aggregate the partial counters before being able to compute the splitting criterion.

Of course, the vertically-parallel design has also its drawbacks. In particular, horizontal parallelism achieves a good load balance more easily, even though solutions for these problems have recently been proposed [20,21]. In addition, if the instance stream arrives in row-format, it needs to be transformed in column-format, and this transformation generates additional CPU overhead at the source. Indeed, each attribute that constitutes an instance needs to be sent independently, and needs to carry the class label of its instance. Therefore, both the number of messages and the size of the data transferred increase. Nevertheless the advantages of vertical parallelism outweigh its disadvantages for several real-world settings.

6.2   Algorithm Structure

We are now ready to explain the structure of the VHT algorithm. In general, there are two main parts to the Hoeffding tree algorithm: *sorting* the instances through the current model, and accumulating *statistics* of the stream at each leaf node. This separation offers a neat cut point to modularize the algorithm in two separate components. We call the first component *model aggregator*, and the second component *local statistics*. Figure 2 presents a visual depiction of the algorithm, specifically, of its components and of how the data flow among them.

The model aggregator holds the current model (the tree) produced so far in a Processor Instance node (PI). Its main duty is to receive the incoming instances and sort them to the correct leaf. If the instance is unlabeled, the model predicts the label at the leaf and sends it downstream (e.g., for evaluation). Otherwise, if the instance is labeled it is used as training data. The VHT decomposes the

---

Algorithm 1. Model Aggregator: VerticalHoeffdingTreeInduction(E, $VHT\_$ *tree*)

Require: E is a training instance from source PI, wrapped in **instance** content event
Require: VHT_tree is the current state of the decision tree in model-aggregator PI
1: Use VHT_tree to sort E into a leaf l
2: Send **attribute** content events to local-statistic PIs
3: Increment the number of instances seen at l (which is n_l)
4: **if** $n_l \bmod n_{min} = 0$ **and** not all instances seen at l belong to the same class **then**
5:     Add l into the list of splitting leaves
6:     Send **compute** content event with the id of leaf l to all local-statistic PIs
7: **end if**

Algorithm 2. Local Statistic: UpdateLocalStatistic(*attribute*, *local_statistic*)

Require: *attribute* is an **attribute** content event
Require: *local_statistic* is the local statistic, could be implemented as $Table <$ $leaf\_id, attribute\_id >$
1: Update *local_statistic* with data in *attribute*: attribute value, class value and instance weights

---

instance into its constituent attributes, attaches the class label to each, and sends them independently to the following stage, the *local statistics* (independent PIs). Algorithm 1 shows a pseudocode for the model aggregator.

The local statistics contain the sufficient statistics n_{ijk} for a set of attribute-value-class triplets. Conceptually, the local statistics can be viewed as a large distributed table, indexed by leaf id (row), and attribute id (column). The value of the cell represents a set of counters, one for each pair of attribute value and class. The local statistics accumulate statistics on the data sent by the model aggregator. Pseudocode for the update function is shown in Algorithm 2.

In Apache SAMOA, we implement vertical parallelism by connecting the model to the statistics via key grouping. We use a composite key made by the leaf id and the attribute id. Horizontal parallelism can similarly be implemented via shuffle grouping on the instances themselves.

Leaf Splitting. Periodically, the model aggregator will try to see if the model needs to evolve by splitting a leaf. When a sufficient number of instances n_{min} have been sorted through a leaf, and not all instances that reached l belong to the same class (line 4, Algorithm 1), the aggregator will send a broadcast message to the statistics, asking to compute the split criterion for the given leaf id. The statistics will get the table corresponding to the leaf, and for each attribute compute the splitting criterion in parallel (an information-theoretic function such as information gain or entropy). Each local statistic will then send back to the model the top two attributes according to the chosen criterion, together with their scores $(\overline{G}_l(X_i^{local}), i = a, b;$ Algorithm 3).

Subsequently, the model aggregator (Algorithm 4) simply needs to compute the overall top two attributes received so far from the available statistics, apply

Algorithm 3. Local Statistic: ReceiveComputeMessage(*compute*, *local_statistic*)

Require: *compute* is an `compute` content event
Require: *local_statistic* is the local statistic, could be implemented as *Table < leaf_id, attribute_id >*
1: Get leaf l ID from `compute` content event
2: For each attribute i that belongs to leaf l in local statistic, compute $\overline{G}_l(X_i)$
3: Find X_a^{local}, which is the attribute with highest $\overline{G}_l$ based on the local statistic
4: Find X_b^{local}, which is the attribute with second highest $\overline{G}_l$ based on the local statistic
5: Send X_a^{local} and X_b^{local} using `local-result` content event to model-aggregator PI via `computation-result` stream

the Hoeffding bound (line 4), and see whether the leaf needs to be split (line 5). The algorithm also computes the criterion for the scenario where no split takes places $(X_\emptyset)$. In [11], this inclusion of a no-split scenario is referred with the term *pre-pruning*. The decision to split or not is taken after a time has elapsed, as explained next.

By using the top two attributes, the model aggregator computes the difference of their splitting criterion values $\Delta\overline{G}_l = \overline{G}_l(X_a) - \overline{G}_l(X_b)$. To determine whether the leaf needs to be split, it compares the difference $\Delta\overline{G}_l$ to the Hoeffding bound $\epsilon = \sqrt{\frac{R^2 \ln(1/\delta)}{2n_l}}$ for the current confidence parameter δ (where R is the range of possible values of the criterion). If the difference is larger than the bound $(\Delta\overline{G}_l > \epsilon)$, then X_a is the best attribute with high confidence $1 - \delta$, and can therefore be used to split the leaf. If the best attribute is the no-split scenario $(X_\emptyset)$, the algorithm does not perform any split. The algorithm also uses a tie-breaking τ mechanism to handle the case where the difference in splitting criterion between X_a and X_b is very small. If the Hoeffding bound becomes smaller than τ $(\Delta\overline{G}_l < \epsilon < \tau)$, then the current best attribute is chosen regardless of the values of $\Delta\overline{G}_l$.

Two cases can arise: the leaf needs splitting, or it doesn't. In the latter case, the algorithm simply continues without taking any action. In the former case instead, the model modifies the tree by splitting the leaf l on the selected attribute, replacing l with an internal node (line 6), and generating a new leaf for each possible value of the branch (these leaves are initialized using the class distribution observed at the best attribute splitting at l (line 8)). Then, it broadcasts a `drop` message containing the former leaf id to the local statistics (line 10). This message is needed to release the resources held by the leaf and make space for the newly created leaves. Subsequently, the tree can resume sorting instances to the new leaves. The local statistics will create a new table for the new leaves lazily, whenever they first receive a previously unseen leaf id. In its simplest version, while the tree adjustment is performed, the algorithm drops the new incoming instances. We show in the next section an optimized version that buffers them to improve accuracy.

Algorithm 4. Model Aggregator: Receive(*local_result*, *VHT_tree*)

Require: *local_result* is an `local-result` content event
Require: *VHT_tree* is the current state of the decision tree in model-aggregator PI
1: Get correct leaf l from the list of splitting leaves
2: Update X_a and X_b in the splitting leaf l with X_a^{local} and X_b^{local} from *local_result*
3: **if** *local_results* from all local-statistic PIs received or time out reached **then**
4:     Compute Hoeffding bound $\epsilon = \sqrt{\frac{R^2 \ln(1/\delta)}{2n_l}}$
5:     **if** $X_a \neq X_\emptyset$ and $(\overline{G}_l(X_a) - \overline{G}_l(X_b) > \epsilon$ **or** $\epsilon < \tau)$ **then**
6:         Replace l with a split-node on X_a
7:         **for all** branches of the split **do**
8:             Add new leaf with derived sufficient statistic from split node
9:         **end for**
10:        Send `drop` content event with id of leaf l to all local-statistic PIs
11:    **end if**
12: **end if**

Messages. During the VHT execution, the type of events being sent and received from the different parts of the algorithm are summarized in Table 1.

Table 1. Different type of content events used during the execution of the VHT algorithm. MA: model aggregator PI, LS: local statistic PI, S: source PI.

Name	Parameters	From	To
`instance;`	<attr 1, ..., attr m, class C>	S	MA
`attribute;`	<attr id, attr value, class C>	MA	LS id = <leaf id + attr id>
`compute;`	<leaf id>	MA	All LS
`local-result;`	<$\overline{G}_l(X_a^{local}), \overline{G}_l(X_b^{local})$>	LS id	MA
`drop;`	<leaf id>	MA	All LS

6.3 Experiments

In our experimental evaluation of the VHT method, we aim to study the following questions:

Q1: How does a centralized VHT compare to a centralized hoeffding tree (available in MOA) with respect to accuracy and throughput?
Q2: How does the vertical parallelism used by VHT compare to horizontal parallelism?
Q3: What is the effect of number and density of attributes?
Q4: How does discarding or buffering instances affect the performance of VHT?

Experimental Setup. In order to study these questions, we experiment with five datasets (two synthetic generators and three real datasets), five different versions of the hoeffding tree algorithm, and up to four levels of computing parallelism. We measure classification accuracy during the execution and at the end, and throughput (number of classified instances per second). We execute each experimental configuration ten times, and report the average of these measures.

Synthetic Datasets. We use synthetic data streams produced by two random generators: one for dense and one for sparse attributes.

- **Dense attributes** are extracted from a random decision tree. We test different number of attributes, and include both categorical and numerical types. The label for each configuration is the number of categorical-numerical used (e.g, 100-100 means the configuration has 100 categorical and 100 numerical attributes). We produce 10 differently seeded streams with 1M instances for each tree, with one of two balanced classes in each instance, and take measurements every 100k instances.
- **Sparse attributes** are extracted from a random tweet generator. We test different dimensionalities for the attribute space: 100, 1k, 10k. These attributes represent the appearance of words from a predefined bag-of-words. On average, the generator produces 15 words per tweet (size of a tweet is Gaussian), and uses a Zipf distribution with skew $z = 1.5$ to select words from the bag. We produce 10 differently seeded streams with 1M tweets in each stream. Each tweet has a binary class chosen uniformly at random, which conditions the Zipf distribution used to generate the words.

Real Datasets. We also test VHT on three real data streams to assess its performance on benchmark data.[7]

- (*elec*) Electricity. This dataset has 45312 instances, 8 numerical attributes and 2 classes.
- (*phy*) Particle Physics. This dataset has 50000 instances for training, 78 numerical attributes and 2 classes.
- (*covtype*) CovertypeNorm. This dataset has 581012 instances, 54 numerical attributes and 7 classes.

Algorithms. We compare the following versions of the hoeffding tree algorithm.

- **MOA:** This is the standard Hoeffding tree in MOA.
- **local:** This algorithm executes VHT in a local, sequential execution engine. All split decisions are made in a sequential manner in the same process, with no communication and feedback delays between statistics and model.
- **wok:** This algorithm discards instances that arrive during a split decision. This version is the vanilla VHT.
- **wk(z):** This algorithm sends instances that arrive during a split decision downstream. In also adds instances to a buffer of size z until full. If the split decision is taken, it replays the instances in the buffer through the new tree model. Otherwise, it discards the buffer, as the instances have already been incorporated in the statistics downstream.

---

[7] http://moa.cms.waikato.ac.nz/datasets/, http://osmot.cs.cornell.edu/kddcup/data sets.html.

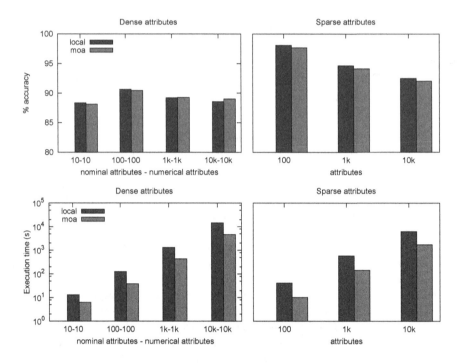

Fig. 3. Accuracy and execution time of VHT executed in local mode on Apache SAMOA compared to MOA, for dense and sparse datasets.

- **sharding:** Splits the incoming stream horizontally among an ensemble of Hoeffding trees. The final prediction is computed by majority voting. This method is an instance of horizontal parallelism applied to Hoeffding trees.

Experimental Configuration. All experiments are performed on a Linux server with 24 cores (Intel Xeon X5650), clocked at 2.67 GHz, L1d cache: 32 kB, L1i cache: 32 kB, L2 cache: 256 kB, L3 cache: 12288 kB, and 65 GB of main memory. On this server, we run a Storm cluster (v0.9.3) and zookeeper (v3.4.6). We use Apache SAMOA v0.4.0 (development version) and MOA v2016.04 available from the respective project websites.

We use several parallelism levels in the range of $p = 2, \ldots, 16$, depending on the experimental configuration. For dense instances, we stop at $p = 8$ due to memory constraints, while for sparse instances we scale up to $p = 16$. We disable model replication (i.e., use a single model aggregator), as in our setup the model is not the bottleneck.

Accuracy and Time of VHT Local vs. MOA. In this first set of experiments, we test if VHT is performing as well as its counterpart hoeffding tree in MOA. This is mostly a sanity check to confirm that the algorithm used to build the VHT does not affect the performance of the tree when all instances are processed

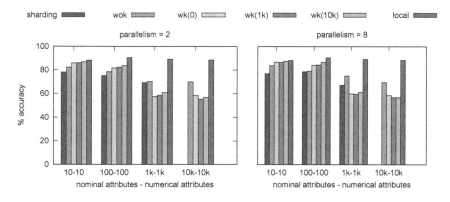

Fig. 4. Accuracy of several versions of VHT (local, **wok**, **wk(z)**) and sharding, for dense datasets.

sequentially by the model. To verify this fact, we execute VHT local and MOA with both dense and sparse instances. Figure 3 shows that VHT local achieves the same accuracy as MOA, even besting it at times. However, VHT local always takes longer than MOA to execute. Indeed, the local execution engine of Apache SAMOA is optimized for simplicity rather than speed. Therefore, the additional overhead required to interface VHT to DSPEs is not amortized by scaling the algorithm out. Future optimized versions of VHT and the local execution engine should be able to close this gap.

Accuracy of VHT Local vs. Distributed. Next, we compare the performance of VHT local with VHT built in a distributed fashion over multiple processors for scalability. We use up to $p = 8$ parallel statistics, due to memory restrictions, as our setup runs on a single machine. In this set of experiments we compare the different versions of VHT, **wok** and **wk(z)**, to understand what is the impact of keeping instances for training after a model's split. Accuracy of the model might be affected, compared to the local execution, due to delays in the feedback loop between statistics and model. That is, instances arriving during a split will be classified using an older version of the model compared to the sequential execution. As our target is a distributed system where independent processes run without coordination, this delay is a characteristic of the algorithm as much as of the distributed SPE we employ.

We expect that buffering instances and replaying them when a split is decided would improve the accuracy of the model. In fact, this is the case for dense instances with a small number of attributes (i.e., around 200), as shown in Fig. 4. However, when the number of available attributes increases significantly, the load imposed on the model seems to outweigh the benefits of keeping the instances for replaying. We conjecture that the increased load in computing the splitting criterion in the statistics further delays the feedback to compute the split. Therefore, a larger number of instances are classified with an older model, thus negatively

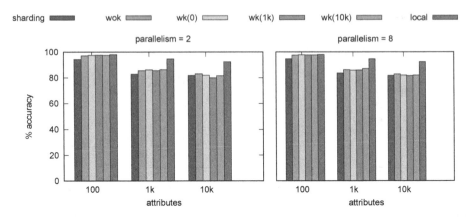

Fig. 5. Accuracy of several versions of VHT (local, **wok**, **wk(z)**) and sharding, for sparse datasets.

affecting the accuracy of the tree. In this case, the additional load imposed by replaying the buffer further delays the split decision. For this reason, the accuracy for VHT **wk(z)** drops by about 30% compared to VHT local. Conversely, the accuracy of VHT **wok** drops more gracefully, and is always within 18% of the local version.

VHT always performs approximatively 10% better than sharding. For dense instances with a large number of attributes (20k), sharding fails to complete due to its memory requirements exceeding the available memory. Indeed, sharding builds a full model for each shard, on a subset of the stream. Therefore, its memory requirements are p times higher than a standard hoeffding tree.

When using sparse instances, the number of attributes per instance is constant, while the dimensionality of the attribute space increases. In this scenario, increasing the number of attributes does not put additional load on the system. Indeed, Fig. 5 shows that the accuracy of all versions is quite similar, and close to the local one. This observation is in line with our conjecture that the overload on the system is the cause for the drop in accuracy on dense instances.

We also study how the accuracy evolves over time. In general, the accuracy of all algorithms is rather stable, as shown in Figs. 6 and 7. For instances with 10 to 100 attributes, all algorithms perform similarly. For dense instances, the versions of VHT with buffering, **wk(z)**, outperform **wok**, which in turn outperforms sharding. This result confirms that buffering is beneficial for small number of attributes. When the number of attributes increases to a few thousand per instance, the performance of these more elaborate algorithms drops considerably. However, the VHT **wok** continues to perform relatively well and better than sharding. This performance, coupled with good speedup over MOA (as shown next) makes it a viable option for streams with a large number of attributes and a large number of instances.

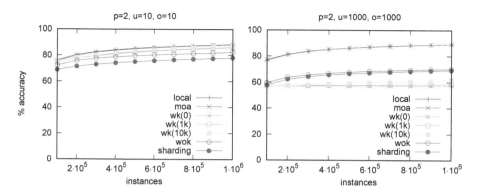

Fig. 6. Evolution of accuracy with respect to instances arriving, for several versions of VHT (local, **wok**, **wk(z)**) and sharding, for dense datasets.

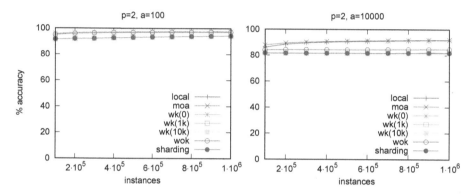

Fig. 7. Evolution of accuracy with respect to instances arriving, for several versions of VHT (local, **wok**, **wk(z)**) and sharding, for sparse datasets.

Speedup of VHT Distributed vs. MOA. Since the accuracy of VHT **wk(z)** is not satisfactory for both types of instances, next we focus our investigation on VHT **wok**.

Figure 8 shows the speedup of VHT for dense instances. VHT **wok** is about 2–10 times faster than VHT local and up to 4 times faster than MOA. Clearly, the algorithm achieves a higher speedup when more attributes are present in each instance, as (*i*) there is more opportunity for parallelization, and (*ii*) the implicit load shedding caused by discarding instances during splits has a larger effect. Even though sharding performs well in speedup with respect to MOA on small number of attributes, it fails to build a model for large number of attributes due to running out of memory. In addition, even for a small number of attributes, VHT **wok** outperforms sharding with a parallelism of 8. Thus, it is clear from the results that the vertical parallelism used by VHT offers better scaling behavior than the horizontal parallelism used by sharding.

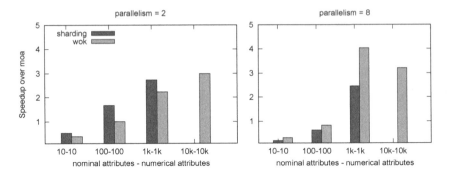

Fig. 8. Speedup of VHT **wok** executed on Apache SAMOA compared to MOA for dense datasets.

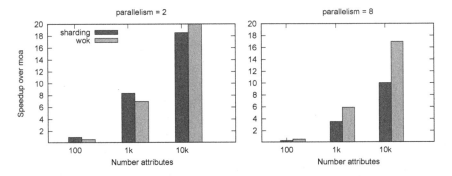

Fig. 9. Speedup of VHT **wok** executed on Apache SAMOA compared to MOA for sparse datasets.

When testing the algorithms on sparse instances, as shown in Fig. 9, we notice that VHT **wok** can reach up to 60 times the throughput of VHT local and 20 times the one of MOA (for clarity we only show the results with respect to MOA). Similarly to what observed for dense instances, a higher speedup is observed when a larger number of attributes are present for the model to process. This very large superlinear speedup ($20\times$ with $p = 2$), is due to the aggressive load shedding implicit in the **wok** version of VHT. The algorithm actually performs consistently less work than the local version and MOA.

However, note that for sparse instances the algorithm processes a constant number of attributes, albeit from an increasingly larger space. Therefore, in this setup, **wok** has a constant overhead for processing each sparse instance, differently from the dense case. VHT **wok** outperforms sharding in most scenarios and especially for larger numbers of attributes and larger parallelism.

Increased parallelism does not impact accuracy of the model (see Figs. 4 and 5), but its throughput is improved. Boosting the parallelism from 2 to 4 makes VHT **wok** up to 2 times faster. However, adding more processors does not improve speedup, and in some cases there is a slowdown due to additional communication overhead (for dense instances). Particularly for sparse instances, parallelism does not impact accuracy which enables handling large sparse data streams while achieving high speedup over MOA.

6.4 Performance on Real-World Datasets

Tables 2 and 3 show the performance of VHT, either running in a local mode or in a distributed fashion over a storm cluster of a few processors. We also test two different versions of VHT: **wok** and wk(0). In the same tables we compare VHT's performance with MOA and sharding.

The results from these real datasets demonstrate that VHT can perform similarly to MOA with respect to accuracy and at the same time process the instances faster. In fact, for the larger dataset, covtypeNorm, VHT **wok** exhibits 1.8 speedup with respect to MOA, even though the number of attributes is not very large (54 numeric attributes). VHT **wok** also performs better than sharding, even though the latter is faster in some cases. However, the speedup offered by sharding decreases when the parallelism level is increased from 2 to 4 shards.

Table 2. Average accuracy (%) for different algorithms, with parallelism level (p), on the real-world datasets.

Dataset	MOA	VHT					Sharding	
		local	**wok**	**wok**	wk(0)	wk(0)		
			p=2	p=4	p=2	p=4	p=2	p=4
elec	75.4	75.4	75.0	75.2	75.4	75.6	74.7	74.3
phy	63.3	63.8	62.6	62.7	63.8	63.7	62.4	61.4
covtype	67.9	68.4	68.0	68.8	67.5	68.0	67.9	60.0

Table 3. Average execution time (seconds) for different algorithms, with parallelism level (p), on the real-world datasets.

Dataset	MOA	VHT					Sharding	
		local	**wok**	**wok**	wk(0)	wk(0)		
			p=2	p=4	p=2	p=4	p=2	p=4
elec	1.09	1	2	2	2	2	2	2.33
phy	5.41	4	3.25	4	3	3.75	3	4
covtype	21.77	16	12	12	13	12	9	11

6.5 Summary

In conclusion, our VHT algorithm has the following performance traits. We learned that for a small number of attributes, it helps to buffer incoming instances that can be used in future decisions of split. For larger number of attributes, the load in the model can be high and larger delays can be observed in the integration of the feedback from the local statistics into the model. In this case, buffered instances may not be used on the most up-to-date model and this can penalize the overall accuracy of the model.

With respect to a centralized sequential tree model (MOA), it processes dense instances with thousands of attributes up to 4× faster with only 10−20% drop in accuracy. It can also process sparse instances with thousands of attributes up to 20× faster with only 5−10% drop in accuracy. Also, its ability to build the tree in a distributed fashion using tens of processors allows it to scale and accommodate thousands of attributes and parse millions of instances. Competing methods cannot handle these data sizes due to increased memory and computational complexity.

7   Distributed AMRules

Decision rule learning is a category of machine learning algorithms whose goal is to extract a set of decision rules from the training data. These rules are later used to predict the unknown *label* values for test data. A rule is a logic expression of the form:

IF *antecedent* **THEN** *consequent*

or, equivalently, *head* ← *body*, where *head* and *body* correspond to the *consequent* and *antecedent*, respectively.

The body of a rule is a conjunction of multiple clauses called *features*, each of which is a condition on an attribute of the instances. Such conditions consist of the identity of an attribute, a threshold value and an operator. For instance, the feature "$x < 5$" is a condition on attribute x, with threshold value 5 and operator *less-than* ($<$). An instance is said to be *covered* by a rule if its attribute values satisfy all the features in the rule body. The head of the rule is a function to be applied on the covered instances to determine the their label values. This function can be a constant or a function of the attributes of the instances, e.g.,

$$ax + b \leftarrow x < 5$$

AMRules is an algorithm for learning regression rules on streaming data. It incrementally constructs the rule model from the incoming data stream. The rule model consists of a set of *normal* rules (which is empty at the beginning), and a *default* rule. Each normal rule is composed of 3 parts: a *body* which is a list of features, a *head* with information to compute the prediction for those instance covered by the rule, and *statistics* of past instances to decide when and how to add a new feature to its body. The default rule is a rule with an empty *body*.

For each incoming instance, AMRules searches the current rule set for those rules that cover the instance. If an instance is not covered by any rule in the set, it is considered as being covered by the default rule. The heads of the rules

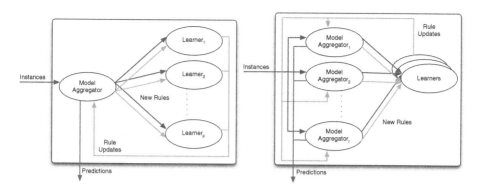

Fig. 10. (Left) Vertical AMRules (VAMR). (Right) AMRules with multiple horizontally parallelized Model Aggregators.

are first used to compute the prediction for the instance they cover. Later, their statistics are updated with the attribute values and label value of the instance. There are two possible modes of operation: ordered and unordered. In ordered-rules mode, the rules are checked according to the order of their creation, and only the first rule is used for prediction and then updated. In unordered-rules mode, all covering rules are used and updated. In this work, we focus on the former which is more often used albeit more challenging to parallelize.

Each rule tries to expand its body after it receives N_m updates. In order to decide on the feature to expand, the rule incrementally computes a standard deviation reduction (SDR) measure [14] for each potential feature. Then, it computes the *ratio* of the second-largest SDR value over the largest SDR value. This *ratio* is used with a high confidence interval ϵ computed using the Hoeffding bound [13] to decide to expand the rule or not: if $ratio + \epsilon < 1$, the rule is expanded with the feature corresponding to the largest SDR value. Besides, to avoid missing a good feature when there are two (or more) equally good ones, rules are also expanded if the Hoeffding bound ϵ falls below a threshold. If the default rule is expanded, it becomes a *normal* rule and is added to the rule set. A new default rule is initialized to replace the previous one.

Each rule records its prediction error and applies a modified version of the Page-Hinkley test [18] for streaming data to detect changes. If the test indicates that the cumulative error has exceeded a threshold, the rule is evicted from the rule set. The algorithm also employs outlier detection to check if an instance, though being covered by a rule, is an anomaly. If an instance is deemed as such, it is treated as if the rule does not cover it and is checked against other rules.

This section describes two possible strategies to parallelize AMRules that are implemented in Apache SAMOA.

7.1 Vertical Parallelism

In AMRules, each rule can evolve independently, as its expansion is based solely on the statistics of instances it covers. Also, searching for the best feature among all possible ones in an attempt to expand a rule is computationally expensive.

Given these observations, we decide to parallelize AMRules by delegating the training process of rules to multiple *learner processors*, each of which handles only a subset of the rules. Besides the learners, a *model aggregator processor* is also required to filter and redirect the incoming instances to the correct learners. The aggregator manages a set of simplified rules that have only head and body, i.e., do not keep statistics. The bodies are used to identify the rules that cover an instance, while the heads are used to compute the prediction. Each instance is forwarded to the designated learners by using the ID of the covering rule. At the learners, the corresponding rules' statistics are updated with the forwarded instance. This parallelization scheme guarantees that the rules created are the same as in the sequential algorithm. Figure 10(left) depicts the design of this vertically parallelized version of AMRules, or Vertical AMRules (VAMR for brevity).

The model aggregator also manages the statistics of the default rule, and updates it with instances not being covered by any other rule in the set. When the default rule is expanded and adds a new rule to the set, the model aggregator sends a message with the newly added rule to one of the learners, which will be responsible for its management. The assignment of a rule to a learner is done based on the rule's ID. All subsequent instances that are covered by this rule are forwarded to the same learner.

At the same time, learners update the statistics of each corresponding rule with each processed instance. When enough statistics have been accumulated and a rule is expanded, the new feature is sent to the model aggregator to update the body of the rule. Learners can also detect changes and remove existing rules. In such an event, learners will inform the model aggregator with a message containing the removed rule ID.

As each rule is replicated in the model aggregator and in one of the learners, their bodies in model aggregator might not be up-to-date. The delay between rule expansion in the learner and model update in the aggregator depends mainly on the queue length at the model aggregator. The queue length, in turn, is proportional to the volume and speed of the incoming data stream. Therefore, instances that are in the queue before the model update event might be forwarded to a recently expanded rule which no longer covers the instance.

Coverage test is performed again at the learner, thus the instance is dropped if it was incorrectly forwarded. Given this additional test, and given that rule expansion can only increase the selectivity of a rule, when using unordered rules the accuracy of the algorithm is unaltered. However, in ordered-rules mode, these temporary inconsistencies might affect the statistics of other rules because the instance should have been forwarded to a different rule.

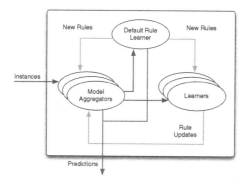

Fig. 11. Hybrid AMRules (HAMR) with multiple Model Aggregators and separate Default Rule Learner.

7.2 Horizontal Parallelism

A bottleneck in VAMR is the centralized model aggregator. Given that there is no straightforward way to vertically parallelize the execution of the model aggregator while maintaining the order of the rules, we explore an alternative based on horizontal parallelism. Specifically, we introduce multiple replicas of the model aggregator, so that each replica maintains the same copy of the rule set but processes only a portion of the incoming instances.

Horizontally Parallelized Model Aggregator. The design of this scheme is illustrated in Fig. 10(right). The basic idea is to extend VAMR and accommodate multiple model aggregators into the design. Each model aggregator still has a rule set and a default rule. The behavior of this scheme is similar to VAMR, except that each model aggregator now processes only a portion of the input data, i.e., the amount of instances each of them receives is inversely proportional to the number of model aggregators. This will affect the prediction statistics and, most importantly, the training statistics of the default rules.

Since each model aggregator processes only a portion of the input stream, each default rule is trained independently with different portions of the stream. Thus, these default rules will evolve independently and potentially create overlapping or conflicting rules. This fact also introduces the need for a scheme to synchronize and order the rules created by different model aggregators. Additionally, at the beginning, the scheme is less reactive compared to VAMR as it requires more instances for the default rules to start expanding. Besides, as the prediction function of each rule is adaptively constructed based on attribute values and label values of past instances, having only a portion of the data stream will lead to having less information and potentially lower accuracy. We show how to address these issues next.

Centralized Rule Creation. In order to address the issues with distributed creation of rules, we move the default rule in model aggregators to a specialized default rule learner processor. With the introduction of this new component, some modifications are required in the model aggregators, but the behavior of the learners is still the same as in VAMR. However, as a result, all the model aggregators will be in synch.

As the default rule is now moved to the designated learner, those instances that are not covered by any rules are forwarded from the model aggregators to this learner. This specialized learner updates its statistics with the received instances and, when the default rule expands, it broadcasts the newly created rule to the model aggregators. The new rule is also sent to the assigned learner, as determined by the rule's ID. The components of this scheme are shown in Fig. 11, where this scheme is referred as Hybrid AMRules (HAMR), as it is a combination of vertical and horizontal parallelism strategies.

8   Conclusions

We presented the Apache SAMOA platform for mining big data streams. The platform supports the most common machine learning tasks such as classification, clustering, and regression. It also provides a simplified API for developers that allows to implement distributed streaming algorithms easily.

Apache SAMOA is already available and can be found online at https://samoa.incubator.apache.org/. The website includes a wiki, an API reference, and a developer's manual. Several examples of how the software can be used are also available. The code is hosted on GitHub. Apache SAMOA contains a test suite that is run on each commit on the GitHub repository via a continuous integration server. Finally, Apache SAMOA is released as open source software under the Apache Software License v2.0.

References

1. Aggarwal, C.C.: Data Streams: Models and Algorithms. Springer, New York (2007). https://doi.org/10.1007/978-0-387-47534-9
2. Ben-Haim, Y., Tom-Tov, E.: A streaming parallel decision tree algorithm. JMLR **11**, 849–872 (2010)
3. Bifet, A., Holmes, G., Kirkby, R., Pfahringer, B.: MOA: massive online analysis. J. Mach. Learn. Res. **11**, 1601–1604 (2010)
4. Bordino, I., Kourtellis, N., Laptev, N., Billawala, Y.: Stock trade volume prediction with Yahoo finance user browsing behavior. In: 30th International Conference on Data Engineering (ICDE), pp. 1168–1173. IEEE (2014)
5. Chatzakou, D., Kourtellis, N., Blackburn, J., De Cristofaro, E., Stringhini, G., Vakali, A.: Mean birds: detecting aggression and bullying on Twitter. In: 9th International Conference on Web Science (WebSci). ACM (2017)
6. Chen, C., Zhang, J., Chen, X., Xiang, Y., Zhou, W.: 6 million spam tweets: a large ground truth for timely Twitter spam detection. In International Conference on Communications (ICC). IEEE (2015)
7. De Francisci Morales, G.: SAMOA: a platform for mining big data streams. In: RAMSS 2013: 2nd International Workshop on Real-Time Analysis and Mining of Social Streams @WWW 2013 (2013)

8. De Francisci Morales, G., Bifet, A.: SAMOA: scalable advanced massive online analysis. JMLR J. Mach. Learn. Res. **16**, 149–153 (2014)
9. Dean, J., Ghemawat, S.: MapReduce: simplified data processing on large clusters. In OSDI 2004: 6th Symposium on Operating Systems Design and Implementation, pp. 137–150. USENIX Association (2004)
10. Devooght, R., Kourtellis, N., Mantrach, A.: Dynamic matrix factorization with priors on unknown values. In: 21st International Conference on Knowledge Discovery and Data Mining (SIGKDD), pp. 189–198. ACM (2015)
11. Domingos, P., Hulten, G.: Mining high-speed data streams. In: KDD 2000: 6th International Conference on Knowledge Discovery and Data Mining, pp. 71–80 (2000)
12. Gama, J., Zliobaite, I., Bifet, A., Pechenizkiy, M., Bouchachia, A.: A survey on concept drift adaptation. ACM Comput. Surv. **46**(4), 44 (2014)
13. Hoeffding, W.: Probability inequalities for sums of bounded random variables. J. Am. Stat. Assoc. **58**(301), 13–30 (1963)
14. Ikonomovska, E., Gama, J., Džeroski, S.: Learning model trees from evolving data streams. Data Min. Knowl. Discov. **23**(1), 128–168 (2011)
15. Jacobs, A.: The pathologies of big data. Commun. ACM **52**(8), 36–44 (2009)
16. Kourtellis, N., Bonchi, F., De Francisci Morales, G.: Scalable online betweenness centrality in evolving graphs. IEEE Trans. Knowl. Data Eng. **27**, 2494–2506 (2015)
17. Kourtellis, N., De Francisci Morales, G., Bifet, A.: VHT: vertical Hoeffding tree. In BigData 2016: 4th IEEE International Conference on Big Data (2016)
18. Page, E.: Continuous inspection schemes. Biometrika **41**, 100–115 (1954)
19. Thu Vu, A., De Francisci Morales, G., Gama, J., Bifet, A.: Distributed adaptive model rules for mining big data streams. In: BigData 2014: Second IEEE International Conference on Big Data (2014)
20. Uddin Nasir, M.A., De Francisci Morales, G., Garcia-Soriano, D., Kourtellis, N., Serafini, M.: The power of both choices: practical load balancing for distributed stream processing engines. In: ICDE 2015: 31st International Conference on Data Engineering. IEEE (2015)
21. Uddin Nasir, M.A., De Francisci Morales, G., Kourtellis, N., Serafini, M.: When two choices are not enough: balancing at scale in distributed stream processing. In: ICDE 2016: 32nd International Conference on Data Engineering. IEEE (2016)

Multimodal Behavioral Mobility Pattern Mining and Analysis Using Topic Modeling on GPS Data

Sebastiaan Merino and Martin Atzmueller[✉]

Tilburg University, Tilburg, The Netherlands
{s.m.merino-norambuena,m.atzmuller}@uvt.nl

Abstract. Identifying risky driving behavior is of central importance for increasing traffic safety. This paper tackles the task of analyzing real (naturalistic) driving data captured by in-vehicle sensors using interpretable data science methods. In particular, we focus on symbolic time-series abstraction and the subsequent behavioral profile identification using topic modeling approaches. For our experiments, we utilize a real-world dataset. Our results indicate interesting behavioral driving profiles including important patterns and factors for traffic safety modeling.

1 Introduction

An increase in road accidents has become a central issue in order to identify risky driving behavior, which increases the chance of road fatalities. In order to provide more detailed insights into real-life driving behavior, sensor-based data science is an important emerging research area, i.e., for diminishing the trend in road accidents, and to enhance overall traffic safety, e.g., [21,27,37].

In this paper, an adapted and significantly extended revision of [28], we investigate the construction and analysis of topic models which describe naturalistic driving behavior. The topic models are constructed using time-series of in-vehicle sensor data. By combining interpretable symbolic abstraction methods with appropriate (topic) modeling approaches, interesting driver profiles, as well as interesting behavioral patterns on driving data can be detected. Our contributions can be summarized as follows: (1) We apply a time-series abstraction method (SAX, Symbolic Aggregate Approximation) that enables interpretable elements to be used in the topic models, and (2) we investigate more complex multi-modal behavioural patterns; (3) with that, we propose an interpretable analysis approach enhancing transparency and explainability in the scope of *explicative data mining* [5,6]. For our experiments, we utilize a dataset collected using in-vehicle sensors on naturalistic driving data. Our results indicate interesting driving profiles, as well as behavioral driving patterns which are interpretable given the symbolic data representation, also considering an analysis of alert vs. relatively unalert drivers.

The rest of the paper is structured as follows. Section 2 discusses related work. After that, Sect. 3 details the applied methods. Next, Sect. 4 presents and

© Springer Nature Switzerland AG 2019
M. Atzmueller et al. (Eds.): MUSE 2015/MSM 2015/MSM 2016, LNAI 11406, pp. 68–88, 2019.
https://doi.org/10.1007/978-3-030-34407-8_4

discusses our results. Finally, Sect. 6 concludes with a summary and interesting directions for future work.

2   Related Work

Extensive research has already provided many insights in the field of road safety. For example, [36] mentioned in their study that an increase in motorization has lead to "severe traffic-related causalities" [36, p. 34]. As density in traffic increases, traffic safety is influenced negatively as it leads to an increase in lane changing and overtaking cars [35]. More specifically, [16] mentioned lane changing and overtaking cars has negative effects on traffic safety.

2.1   Profiling Driving Behavior

Besides traffic density having an influence on traffic safety, research has also been conducted to study driving behavior [11,18,19]. Studies included self-reports of large populations to study the behavior of participants while driving a car. The focus in these studies was to examine driving behavior by analyzing risks that have a negative influence such as, drowsiness, intoxication, aggression, or distraction. While it is indisputable that well designed questionnaires allow researchers to test hypotheses, [34] concluded that overconfidence might lead to implications in results of studies which utilize questionnaires to determine driver safety. A reason for this is the "belief that one possesses a greater competence than one's peers" [34, p. 265]. Overestimation, has been known to cause the illusion of control where an individual, who is in an adverse state of driving, might perceive him- or herself in a controllable state. Studies revealed that while individuals perceived themselves in a controllable state, cognitive tasks were performed significantly worse. Thus, overestimation leads to underestimation of risks, which makes individuals not take precautions such as resting, hands-free driving, or not making use of a mobile phone while driving [30].

2.2   Naturalistic Driving Data Abstraction and Modeling

In a study conducted by [27], the authors highlighted the importance of naturalistic driving data (NDD) in studying driving behavior. However, the authors mentioned NDD-analysis is challenging as "the large amount of data commonly collected during naturalistic driving studies makes comprehensive analysis prohibitive without some type of data reduction" [27, p. 2107]. A possible consequence of this reduction might lead to omit subtle information or even the broader context of a data set. Moreover, a challenge to analyze large amounts of data is the interpretation of continuous data. Therefore, [27] included a symbolic representation of time series data by applying Symbolic Aggregate Approximation (SAX) [25], which transparently retains the data characteristics. The conversion of time series data is executed by transforming continuous instances to alphabetical representations. Time series output is normalized and divided into equal sized ranges. Then, each range is presented by a letter, so instances which coincide in a range are assigned to a letter. By applying this transformation to time series, with multiple variables, a *bag-of-words*-model is created. The method of [27] allowed consecutive studies not only to analyze large amounts of time series data, but also to combine various variables [27,29,32]: in the study of [29], traffic data was combined with weather data, while [32] were able to analyze expected events during driving situations of individuals.

The SAX representation of time series data enables extended analysis: More specifically, Probabilistic Topic Modelling (PTM) makes it possible to apply unsupervised learning for exploration of the data, cf. [12]. Intuitively, the output of PTM is a topic model which captures occurrence distributions in text to a set of words. The probability distributions of these sets of words are then assigned to as topics. As documents may contain multiple topics, PTM allows to classify topics in documents. This method has successfully been applied in time series data sets, as it has the potential to analyze large data sets while producing a comprehensive set of topics and find subtle patterns. For example, differences in expected and unexpected events were measured (e.g., involving crosswinds) by including variables into a modeling approach using PTM such as, steering angle and changes in the frequency of steering maneuvers [32]. Besides that, the researchers were able to translate continuous variables to a verbal representation, which was necessary as they applied Latent Dirichlet Allocation (LDA) [12,13] for probabilistic topic modeling in their study.

2.3 Psychomotor Vigilance Test

In order to investigate differences in driving behavior between alert and unalert individuals, the alertness state of participants in a driving simulation needs to be estimated. Multiple studies have included the *Psychomotor Vigilance Test* (PVT) [2,15]. PVT is a reaction time test in which individuals are randomly exposed to stimuli over an period of time ranging from 3 to 10 min per test. During a test, an inter-stimulus appears on a screen which irregularly is displayed between two and ten seconds, and participants are requested to react as fast as possible to the stimulus.

The objective of studies which implemented PVT is to measure whether a decrease in reaction time can be estimated, and thus, depreciation of psychomotor skills. Studies have successfully found differences in reaction times as errors increased and vigilance decreased, e.g., as a result of sleep deprivation [26].

Originally, PVT required to include ten minute intervals per test to ensure validity of alertness results, however five minute intervals in PVT led to similar results as to ten minute intervals [24]. A major advantage of lower intervals is the practicality of conducting experiments with PVT in situations were surroundings are hectic. Besides improvements in time intervals, studies have also measured the effect of conducting PVT on different devices (e.g., mobile phones, tablets). In the past, PVT required to be conducted on a computer. In [1] the difference between computer-based PVT and touchscreen devices was studied, concluding that the results for both tests had no significant differences. Therefore, their study contributed to the validity of hand-held devices with touchscreen, and allowed future studies to apply PVT in more hectic surroundings.

3   Methods

This section establishes the methodology of the current study. We first summarize the collected dataset before we present the modeling method in detail.

3.1   Data Collection

Naturalistic driving data for this study were collected by conducting twenty-five real-life experiments. In collaboration with Crossyn Automotive B.V.[1] in the Netherlands, this study made use of an in-vehicle sensor, which recorded rides during experiments providing GPS sensor data. The GPS data made it possible to measure linear acceleration and speed. In total, twenty-five participants were recruited through convenience sampling, where that group consisted of 5 women, 20 men, with $\mu_{age} = 28.38$, $\sigma_{age} = 8.42$, age range 19–59.[2] Attendees were invited to participate by signing a consent form, data collection was then anonymized. Participants were instructed that they would receive oral directions throughout the driving task, meaning they did not have to navigate themselves. Visual navigation was excluded from the experiment, as drivers were required to stay focused on the road to ensure safety. Also, auditory input in the vehicle was diminished by keeping the radio mute.[3] Finally, participants were allowed to have a casual conversation during the experiment in order to imitate naturalistic driving behavior. Most of the routes showed similarities, covering approximately 30 min of driving on long uneventful roads, where the speed limits vary from 50 km/h to 70 km/h.

3.2   Preprocessing and Feature Extraction

The final dataset consists of 24 rides: One experiment/ride was excluded from the dataset, as the in-vehicle sensor lost its connection due to an unknown reason. In total, the experiment stored 813.30 min of driving data, which equals 13.55 h. Every experimental task took approximately 30 min per ride (MPR, $M_{ride} = 33.89$ MPR, $SD_{ride} = 8.33$ MPR).

The data processing using SAX on the collected time series data was achieved by creating segments of the value range of the respective time series in the data, in which each segment was represented by an alphabetical letter. Then, the alphabet of SAX letters needed to be determined. Referring to the study of [27] in which nine groups were defined, this study applied a similar amount of symbols. However, after conversion of speed to SAX-letters, eight letters could be

---

[1] https://www.crossyn.com/crossyn.

[2] μ = mean, σ = standard deviation.

[3] In their study, [14] concluded that hostile music can lead to distracted drivers.

generated at most. Therefore, the normalized data were divided into eight scales for the attribute speed and nine for the attribute (linear) acceleration. Finally, the letters in the alphabet were represented by nine unique letters. Table 1 illustrates the alphabetical representation of letters used for this study, and the corresponding value ranges that these SAX-letters represented.

Table 1. Conversion of continuous input to SAX output.

SAX letters	Range			
	Speed (km/h)	N	Acceleration (g)	N
A/a	0.0 to 1.0	10566	−1.1890 to 0.0760	3728
B/b	2.0 to 16.0	4161	−0.0708 to −0.0472	2474
C/c	17.0 to 28.0	4407	−0.0437 to −0.0283	6224
D/d	28.7 to 38.0	5417	−0.0212 to −0.0094	56
E/e	39.0 to 48.1	10513	−0.0081 to 0.0071	24230
F/f	48.4 to 59.3	4011	0.0089 to 0.0239	62
G/g	60.0 to 74.0	3054	0.0283 to 0.0438	5208
H/h	75.0 to 124.0	6669	0.0453 to 0.0708	3000
i	n.a.	n.a.	0.0755 to 1.2180	3815

After defining the alphabet, letters were combined into words. The structure of each word consisted of one uppercase letter (i.e. from "A" to "H") representing speed. Subsequently, the second (lowercase) letter in the converted words denotes an acceleration-letter from "a" to "i". Lastly, the letters were joined by placing an underscore between letters of each word. In total, 67 unique words were shaped based on the combinations of speed, and acceleration letters of the dataset.

The current study exhibits similarities with the frequency of speed letters in the study of [27], partially reproducing their results, but in another context, since we aim at identifying distinctive behavioral profiles in a general setting. As described above, for readability purposes, letters which present speed intervals are mentioned by an uppercase (capital) letter and acceleration intervals by lowercase letters. Table 1 already indicated that the most occurring letters in the data set were "A" ($N = 10556$), "E" ($N = 10513$), and "H" ($N = 6669$). As "A" stands for a very low speed, it represents the car being in a stationary position. The letter "E" in its turn is represented by a speed between 39.0 and 48.1 km/h, which results in a modest speed. This speed coincides with maximum speed barriers between 30 to 50 km/h, and can be defined as city driving, cf. [27]. Followed by the "E", the letters "F", and "G" range between speeds of 48.4 to 74.0 km/h, which represents roads where the speed limit is 70 km/h. Henceforth, this driving behavior can be defined as moderate. The letter "H" ranges from 75.0 to 124.0 km/h. This speed range is also similar to the one in the study of [27], in which it was defined as high speed driving. Therefore, the current pre-processing steps indicate similarities to this previous study.

Whereas probabilistic topic modelling applies text documents in order to conduct its analysis, this study makes use of driving data, which typically consists of continuous variables. Therefore, speed and acceleration were first converted to SAX words. Then, since LDA required a document word input for topic modeling, the collection of words in documents was transformed into a *bag-of-words* (BOW) representation, a standard format which is applied in natural language processing, containing the occurring words and their frequencies.

In order to obtain the LDA model, it is required to maximize the log-likelihood [13]. We performed grid search, to determine the topic model with the best fitting topics empirically. The used parameters were: number of topics and learning decay. LDA requests for input for the amount of topics. Therefore, a range from 1 to 5 was given as input. Learning decay is applied to control the learning rate. The values which are set for this range, vary from 0.5 to 1.0. For this grid-search, three inputs were applied (i.e., {0.5, 0.7, 0.9}). The default for learning decay is set at 0.7. The grid search indicated that the best results were obtained with a learning decay of 0.5, yielding three topics.

Unlike the study of [27], in which no words were excluded from the data set, this study included also pre-processing steps which are typically applied in text mining analyses. The following further steps were included in the analysis, as detailed below in Sect. 4:

- n-grams: For the analysis bi-, and trigrams were included.
- n-grams/max-features: This parameter controls the maximum amount of features (e.g., only 100 words) to include in the BOW.
- Selective n-grams: Words which appeared only in one document, or which appeared in more than 50% of the documents, were excluded.

4   Results

In this section, we first focus on probabilistic topic modeling using LDA, reproducing similar results as presented in [27]. After that, we present results on more complex topic models enabling a more comprehensive analysis on complex driving patterns, and analyze those in relation to alert and unalert driver indicators.

4.1   Topic Model Description

The fitted topic model consists of three topics. This section will discuss the topic definition, the topic distribution along the dataset, and a visual representation using the Python package pyLDAvis[4].

Topic Definition. The LDA model has created a topic model of three topics in total. Each topic consists of the probability of keywords which explain the weight of significance. The top keywords are extracted from the topic model

---

[4] https://github.com/bmabey/pyLDAvis/blob/master/pyLDAvis/sklearn.py.

by converting the vectorized dataset to the featured names. This leads to an overview of SAX words, which describe the composition of each topic. Table 2 illustrates the output of each topic by showing the words with the highest weight. Each topic will be further described in the next sections.

Table 2. Topics of top five SAX words with highest weights in descending order.

Words	Topics		
	Topic 1	Topic 2	Topic 3
Word 1	A_e	A_e	H_e
Word 2	E_e	E_e	A_e
Word 3	H_e	E_c	H_c
Word 4	E_g	E_g	H_g
Word 5	G_e	F_e	E_e

Topic 1: City Driving. Topic 1 has the strongest weight for word "A_e", which is characterized by the letter "A" for speed and "e" for acceleration. Previously, Table 1 described the ranges in which each letter coincided, meaning that "A_e" indicates the car in a stationary position. The second strongest word (i.e., "E_e") identifies city driving while accelerating constantly. Third, the letter "H_e" gives, defines a constant higher speed during a ride (i.e., 75.0 and 124.0 km/h) with no acceleration. The fourth word in topic 1 is "E_g", which describes a city driving speed with a higher acceleration. Finally, the word "G_e" describes a higher speed (i.e., 60.0 to 74.0 km/h) and again with no acceleration.

Except for word 4, the top occurring words describe a constant speed between 34.0 and 124.0 km/h. Speed "H" indicates a higher speed, which would occur on highways. At the same time, the lower bound of this range indicates 75 km/h, meaning that roads with speed limits of 70 km/h could occur during a ride. Besides that, the word "A_e" is a frequent occurring word in the dataset, and is listed on top of the other words, indicating that modest speeds are included in this topic, including stopping motions (i.e., stationary position). Thus, topic 1 is described as city driving.

Topic 2: Complete City Driving. Topic 2 is somewhat similar to Topic 1. However, discrepancies are found as indicated by the presence of the words "E_c" and "F_e". First, "E_c" can be described as a speed between 39.0 to 48.1 km/h with deceleration, meaning that speed of the car represents city driving, and is decelerating at the same time. This action will most likely occur right before the car is about to turn into stationary position. The second word, "F_e", which also deviates from Topic 2 is represented by a constant speed between 48.4 and 59.3 km/h.

As Topic 2 differs by the two most occurring words, it resembles city driving similarly to Topic 1. However, the top five words of Topic 2 describe city driving more thoroughly as it includes a speed between 48.4 and 59.3 km/h. The maximum speed of roads which was included during the experiment was mostly 50 km/h. This word is essential in the definition of city driving resulting in a more complete description.

Topic 3: Highway Driving. Topic 3 clearly highlights different top words compared to Topic 1 and 2. The highest ranked word, "H_e", represents a constant speed between 75.0 to 124.0 km/h. Then, Topic 3 includes word "A_e", which is a stationary representation of the car. Third, "H_c" represents a similar speed range as word 1, but instead it denotes a decelerating state. Fourth, "H_g" indicates the same speed range as the word "H_c", but rather than deceleration, this word is a representation of acceleration. Lastly, the word "E_e" is a representation of a constant speed range from 39.0 to 48.1 km/h. As the top words give an indication of higher speeds, this topic can be defined as highway driving.

Topic Representation in Documents. An essential part of LDA analysis is to distinguish which topics belong to which documents. In order to give insights which topic was most dominant in each document, a topic distribution is displayed in Table 3. A surprising result is that 5 rides are clustered as highway driving, which corresponds to the rides that were held in the driving experiment. Thus, LDA analysis succeeded in correctly clustering the documents into the type of driving which was most dominant during each ride.

Table 3. Distribution of dominant topics in current data set.

Topic number	Number of documents
2	19
3	5

Table 4 provides a detailed view on all documents which were included in this study. Then, weights of topics indicate to which extent a topic is present in each document. The weights of all topics, which are included in one document, accumulate to 1.0. In Table 4, Topic 2 and Topic 3 are included, while Topic 1 is excluded. The reason for this is that Topic 1 had no occurrence in any of the rides. Interestingly, LDA analysis found the best fitted model with three clusters, but Topic 1 shows no significance when assigning topics to rides. Below, we highlighted two topics, which in essence appeared to be similar. The results in Table 4 confirm the redundancy of Topic 1 compared to Topic 2 and explain why Topic 1 has no significance in the distribution of topics.

4.2   Visualization of Topics

For visualization, we applied pyLDAvis to the LDA analysis. Figure 1 illustrates the interactive output from the LDA model. On the left, the bubbles represent

Table 4. Weights of topic occurrences per ride (corresponding to the document number) for Topic 2 and 3. Topic 1 is excluded here, since no occurrences were present for this topic in the experimental sample.

Document number	Topic 2	Topic 3
1	0.98	0.02
2	1.00	0.00
3	1.00	0.00
4	1.00	0.00
5	0.34	0.66
6	1.00	0.00
7	1.00	0.00
8	1.00	0.00
9	1.00	0.00
10	1.00	0.00
11	0.86	0.14
12	0.94	0.06
13	0.00	1.00
14	0.00	1.00
15	1.00	0.00
16	0.00	1.00
17	1.00	0.00
18	1.00	0.00
19	1.00	0.00
20	1.00	0.00
21	0.14	0.86
22	1.00	0.00
23	1.00	0.00
24	1.00	0.00

the topics in a semantic topic space. This means that the closer the bubbles are to each other, the more semantic resemblance they share. Figure 1 indicates that Topic 2 and Topic 3 do not share common words, as they appear on a long distance from each other on the distance map.

On the right side of Fig. 1 the words are displayed which were applied to the LDA analysis. The interactive visualization makes it possible to highlight a word. Subsequently, the sizes of the bubbles on the left pane adapt to the prevalence of the word inside the topic, meaning that the higher the importance of a word in a topic, the larger the size of the bubble.

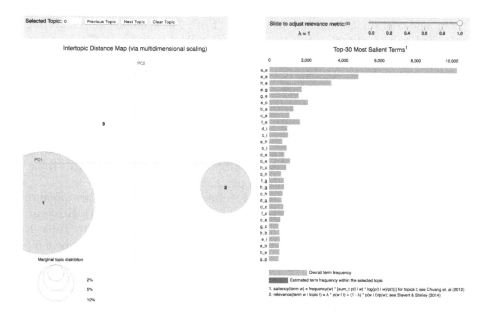

Fig. 1. The topics from the LDA analysis are visualized by applying pyLDAvis. Topics are represented by bubbles on the left side indicating their sizes and respective distances to each other as obtained by multi-dimensional scaling. The right side of the figure displays the word term space, visualizing the respective term frequencies. The words are ranked in descending order of importance.

4.3 Identifying Behavioral Patterns Using Topic Modeling

In the following, we outline the steps for identifying behavioral driving patterns using a more complex topic modeling approach. As we have discussed in Sect. 3, these refinements include using (1) n-grams (bigrams, trigrams), (2) a restricted set of n-grams, and (3) a selection of n-grams forming the topic models.

Behavioral Topic Modeling Using n-Grams: The first experiment applied bi- and trigrams on the data set to enable more complex patterns of words, and thus to potentially find stronger patterns than only applying unigrams. The most optimal log-likelihood was achieved with four topics. An overview of each topic, with most occurring vocabulary is shown in Table 5. In order to investigate if the results of this LDA-model differ from the LDA-model described in Sect. 4.1, we assess each topic individually:

- **Topic 1 (city driving)** forms the largest topic in the LDA-model. The most relevant words indicate similarities with the previously established LDA-model in Sect. 4.1. More specifically, the top bi-grams, which are shown in Fig. 3, are defined as "A_e A_e" and "A_e A_e A_e". As previously explained, "A_e" is a representation of a stationary state of the vehicle. The third and fourth most important words of topic 1 are again a homogeneous combination of bi- and trigram, but instead of the word "E_e". The symbolic representation of speed and acceleration was previously described in Table 1. "E_e" was an indication of a constant moderate speed (39.0 to 48.1 km/h). Compared to Sect. 4.1 topic 1 can be determined as city driving as speeds do not exceed speed letter "H", which is higher than 75.0 km/h.

- **Topic 2 (highway driving)** distinguishes itself, compared to Topic 1, by higher speeds. Section 4.1 defined highway driving as a topic, as speeds were included which were higher than 75.0 km/h. Applying bi- and trigrams indicates that a combination of "H_e" determines the most important word in Topic 2. More specifically, the highest proportion of "H_e" resides this topic. Thus, this LDA-model has defined Topic 2 as highway driving.

- **Topic 3 (city driving with subtle changes)** shows most resemblance with Topic 1 as speeds do not exceed the letter "H" (> 75.0 km/h). However, the word frequencies in Topic 3 are significantly lower compared to Topic 1. However, a more interesting observation is the bigram "E_e D_c", meaning that the car shifted from a constant speed between 39.0 and 48.1 km/h to 28.7 and 38.0 km/h. Moreover, the acceleration decelerates to -0.04 and -0.03. A subsequent event which occurs in Topic 2 is shown with the trigram "E_e D_c D_e", which clearly indicates a change in constant speed from approximately 50 km/h to 30 km/h. Thus, Topic 3 shows a subtle change in speed, which was previously not indicated in Topic 1.

- **Topic 4 (highway driving during rush hour)** is the smallest topic in this LDA-model, as the size in the visualization is minimal. The speed in Topic 4 ranges from minimal (i.e., "A") to a higher speed ("H"). Overall, term frequency is very low in Topic 4, indicating the size of this topic is very small. More interestingly, Topic 4 consists of a combination of high speed, and a stationary position. Referring to the driving experiments of this study, one participant was subject to rush hour, while driving on a highway. As one participant was subject to this situation, the size of Topic 4 is explained.

As implementing the n-grams LDA-model created quite informative topics, we investigated that in more detail. First, we restricted the n-gram set to its top 100 (max-features). This resulted in an optimal LDA-model of four topics. Since that only includes the most common features of a corpus, it is expected that this LDA-model will indicate more general patterns in the driving data.

- **Topic 1 (city driving):** Both aforementioned LDA-models did both include one topic which represented the majority of the data. In this LDA-model, Topic 1 is repeatedly overrepresented. This means, that in the current topic, city driving is represented. All bi- and trigrams, which are included, do not exceed 75.0 km/h. Another observation is that SAX words, which are combined in bi- and trigrams are similar to one another. This indicates that the corpus contains series of driving data which are similar to each other.

- **Topic 2 (highway driving):** Similar to the previous LDA-model, this model included highway driving.

- **Topic 3 (city driving with max 70 km/h)** has the largest representation of "A_e" SAX words. More interestingly, words such as "F_e" and "G_e" are

Table 5. Presentation of most occurring words for each topic in LDA-model with inclusion of bi-, and trigrams.

Rank	Topic vocabulary			
	Topic 1	Topic 2	Topic 3	Topic 4
1	A_e A_e	H_e H_e	A_e A_e A_e	H_e H_e
2	A_e A_e A_e	H_e H_e H_e	A_e A_e	H_e H_e H_e
3	E_e E_e	H_c H_c	E_e E_e	A_e A_e A_e
4	E_e E_e	H_g H_g	F_e F_e	A_e A_e
5	F_e F_e	A_e A_e	E_e E_d	F_e F_e
6	E_c E_c	A_e A_e A_e	F_f F_e	E_e E_e
7	G_e G_e	H_e H_c	F_e E_c	F_e F_e F_e
8	E_g E_g	H_e H_c H_c	E_e D_c	E_e E_e E_e
9	G_e G_e G_e	H_c H_c H_c	E_f E_c	G_e G_e
10	B_a B_a	H_e H_e H_c	E_e E_c E_e	E_c E_c

included which are a representation of speeds higher than 60.0 km/h. This result is an indication of events during the driving experiments where participants were driving on a road with a maximum speed of 70 km/h. The inclusion of stationary words (i.e., "A_e") explain traffic lights which participants encountered on this specific part of the road.

- **Topic 4 (highway driving during rush hour):** Topic 4 is similar to Topic 3, in a sense that higher speeds are combined with the stationary state of the car. In this topic, even higher speeds are recorded. The fact that word combinations with "A_e" exist, indicates a driving situation within a rush hour, as in these driving situations it is common to stand still on a highway due to a high amount of traffic. This result corresponds to one participant, who drove on a highway during rush hour in the afternoon.

Behavioral Topic Modeling Using Selected n-Grams. The last experiment makes it possible to include and exclude features, which are under or over represented in the corpus. The optimal LDA-model in this setting consisted of two topics. Similar to the previous experiments, this experiment included bi- and trigrams. We applied two parameters, i.e., *min-df* and *max-df* for including or excluding features that are underrepresented or overrepresented in the corpus, respectively. The default *min-df* is set at 1, meaning that no features in the corpus are ignored. *max-df* was set at 0.5, meaning that words which occur in 50% of the corpus are removed, to prevent the majority of words to be included. This approach allows to zoom into less frequent words in the corpus, and to focus on subtle changes in driving behavior.

- **Topic 1 (high speed driving):** the largest proportion of this topic consists of combinations, which include the SAX-word "H_e". Moreover, each combination in this topic is defined with the letter "H", which is the representation of speed between 75.0 to 124.0 km/h. Furthermore, the variation of accelerating symbolic representations is almost complete as the acceleration letters "b" to "h" are represented in the topic terms.

- **Topic 2 (high and low acceleration and deceleration):** presents a wider variety of SAX-representations. For example, the first most relevant word in this topic is defined is "B_c B_c B_c", meaning an occurrence during a ride in which a driver would decelerate strongly, while driving a low speed. Typically, in real driving behavior, a state of "B_c" would occur right before the car would reside in stationary position (i.e., "A_e"). The second, most important term in Topic 2 is "C_g C_g C_g".

A closer look at Table 1 reveals that this trigram explains a state in driving behavior in which the car is strongly accelerating, but still in a fairly low speed. The following terms in Topic 2, are indications of high acceleration and deceleration. For example, the bigram "E_c D_b" is an occurrence in which the vehicle is decelerating. On the contrary, "D_h D_i E_i" indicates a strong acceleration from speed "D" to "E". All salient terms in topics of previous experiments, which included bi- and trigrams, and restriction using max-features, indicated mostly zero acceleration with SAX letter "e". The current topic sheds light on SAX-words, which represent a variation of acceleration and deceleration in driving behavior. Since we removed 50% of most occurring bi- and trigrams in the corpus, this time topics were shaped, which include less situations that occurred during the driving experiments.

4.4   Behavioral Topic Profiles: Driver Alertness Analysis

To establish a framework which enables to determine the alertness state of participants, a Psychomotor Vigilance Test (PVT) was applied. The objective of PVT was to objectively quantify the vigilant state of participants prior and after the driving experiment, as it was hypothesized that driving a vehicle would have an effect on individuals.

Collection of PVT Data. During all experiments (i.e., prior and after the driving experiment) participants were exposed to PVT. In order to perform PVT, a mobile application was utilized. Inter-stimuli appeared randomly between 2 and 10 seconds within a time range of 5 min. A visual stimulus, displayed in counting milliseconds, was shown after each inter-stimulus started. Milliseconds continued counting until a participant reacted, by means of simple reaction time, to the stimulus by tapping the touchscreen of the iPad. Simple reaction time is a requested response in experiments, where participants are exposed to one stimulus, and only one reaction is required [31]. Participants who reacted too soon to a inter-stimulus, received immediate feedback, with a brief message (i.e., "False Start"). A *False Start*-indication was displayed in case participants responded before a inter-stimulus appeared up until <100 ms.

Reaction times of <100 ms were previously determined as anticipated reactions of participants [10]. More specifically, [31] determined the average visual reaction time of individuals to be at approximately 209 ms, with a standard deviation of 42.50 ms. The results in the study of [31] did not require to adjust the modality of the current PVT. Reaction times of >500 ms were denoted as *True Errors*, which is a current determination in PVT of errors that occur during attention inducement [10]. In total, 25 × 2 PVT were recorded. Two PVT of one participant were excluded from the dataset, as the data of the driving experiment were not valid for this participant.

The reaction times (RT) in each PVT were divided in *False Start* (i.e., reaction time <100 ms, *FS*), *True Errors* (i.e., reaction time >500 ms, *TE*), and *Correct* (i.e., reaction time ≥100 ms and ≤500 ms, *CC*). Then, the reaction times of each 5-min PVT were prepared for analysis. First, the amount of n FS and n TE were determined. The quantity of errors were distinguished as errors before, and after the driving experiment. Second, the remaining reaction times, denoted as Correct (CC), were grouped per minute to analyze the alteration per minute before, and after the driving experiment.

Data Annotation. In order to categorize the vigilant state of participants, a distinction of (1) *alert*, and *unalert* participants needs to be determined. The objective of annotating participants in one of two states is then to measure if driving conditions have an influence on the vigilant state of participants. Our analysis of the means of the reaction times for the participants before and after the experiment indicated individual differences, however, those were not significant at an individual level. Therefore, we applied an alternative approach, by not only looking at the reaction times themselves, but at the errors made by the individual participants. Here, we made the assumption, that the number of errors correlate with the level of alertness, i.e., that a lower number of errors are observed in an alert state, and higher numbers of errors correspond to an unalert state. In that way, we constructed an abstracted score for the vigilance, as a proxy, which we call the *alertness indicator* as an indication of an alert vs. an unalert state. We then assessed the state of a participant before and after the driving experiment.

Accordingly, in order to annotate whether a participant was affected after the driving experiment, the errors made were compared. An increase in the number of errors (i.e., False Starts and True Errors) indicated an impaired vigilant state, i.e., in decreased alertness indicator. On the contrary, a decrease in errors, indicated an increase in the alertness indicator, which meant that the indicated alertness state of a participant improved after the driving experiment. This approach of labelling was based on the study of [2] in which alertness of participants was based on the amount of errors construction workers made during a 2.5 h experiment. In total, 6 six participants were denoted as having a decreased alertness indicator, versus 18 participants having an increase alertness indicator.

Alertness Indicator Analysis. In the following, we analyze the topic models constructed as described above, concerning the group with alert and unalert indicators, respectively. Concerning the simple LDA models, we actually observed some differences between the two groups – however, these were not large enough to satisfy typical statistical significance criteria when comparing the topic distributions with respect to the features. Therefore, we focus on more complex topic models below, where we consider the individual topic distributions for the two groups regarding alertness in detail.

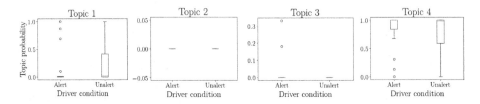

Fig. 2. Mean topic probabilities of alert versus unalert participants. The standard deviations are shown as error bars. The boxplots include the results of the LDA-model, including bi-and trigrams.

Bi-and Trigrams LDA-Model. In this LDA-model, 4 topics were distinguished (i.e., (1) Highway driving, (2) Highway driving during rush hour, (3) City driving with subtle changes, and (4) City driving. The topics of this experiment will be further discussed in this section:

- Topic 1 illustrates the distribution of highway driving. As can be observed in Fig. 2, Topic 1 has a higher probability for alert participants compared to unalert participants. Interestingly, there are four outliers in the alert group. This result resembles the experimental setup in which exactly four participants drove on highways. Furthermore, this can indicate that along the experiments, participants resideded in an alert state, or that participants became more alert as participants went on highways. In addition, the results for unalert participants show that Topic 1 has a low probability, meaning that unalert participants were not handling highway driving.
- As also discussed above, Topic 2 did not occur in the experimental rides.
- Topic 3 indicated city driving with subtle changes. The experimental results indicate that Topic 3 only occurred in 2 rides, and only for alert participants. This indicates that this Topic Model has created an extra topic, as two rides deviated from the majority of the corpus.
- As we illustrated above, Topic 4 formed the largest topic. Here, the results clearly indicate that the weight of Topic 4 is the largest. However, we observe three outliers, which represent alert participants, similar to the outliers for Topic 1, which also resembled highway driving.

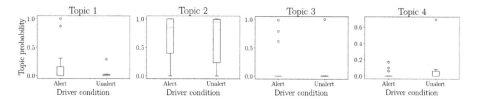

Fig. 3. Mean topic probabilities of alert versus unalert participants. The standard deviations are shown as error bars. The boxplots include the results of the LDA-model, including bi-and trigrams and max-features = 100.

Max-Features Model. As outlined in the previous section above, as the application of bi-, and trigrams resulted in interesting topics, these were included in the next two experimental setups to create LDA-models. First, a *max-features* model was created, which was set at max-features = 100, as we have described in the previous section in more detail. This setup created 4 new topics. The probability of these topics will be further discusses:

- Topic 1 was an indication of a large variety of speeds, which shared resemblance by the acceleration letter *e*, which combined with speed letters, signified constant speeds as acceleration was it its minimum. The results for Topic 1 shown in Fig. 3 indicate a low probability of Topic 1 in both participants groups (i.e., Alert and Unalert participants). Two participants, labelled as alert, contained a high probability of Topic 1, meaning that during their ride, constants speeds were dominant.
- Topic 2, the largest topic in this LDA-model, represented city driving up tot a maximum speed of 75 km/h. The results in the figure illustrate the dominance of the probability of Topic 2 for both types of participants. For both participants groups, the whiskers reach to a probability of 0.0, meaning rides in the corpus did not encounter driving behavior of Topic 2.
- Similar to the previous LDA-model discussed above, the current LDA-model consisted of highway driving. The results in this topic show few participants who were exposed to a highway setting. Figure 3 shows three clear outliers for alert participants, compared to 1 outlier for unalert participants.
- The latter topic, which was previously defined as highway driving during rush hour, is more represented by alert participants, compared to unalert participants. However, unalert participants had one outlier, with a high probability (i.e., topic probability $\pm$ 0.7).

Selected n-Grams Model. The last experimental setup consisted of omitting frequently occurring bi-, and trigrams (i.e., max-df = 0.7) and least occurring (i.e., min-df = 0.1). After applying this setup, 2 topics were extracted from the corpus.

- Figure 4 indicates the results for alert and unalert participants for varying acceleration situations. Alert participants are underrepresented in Topic 1 as the higher bound of the boxplot reaches less than a probability of 0.2.

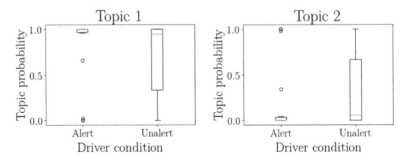

Fig. 4. Mean topic probabilities of alert versus unalert participants. The standard deviations are shown as error bars. The boxplots include the results of the LDA-model, including bi-and trigrams and min-df = 0.1, and max-df = 0.7.

However, two outliers are found in this participant group. Unalert participants are, compared to alert participants, overrepresented in Topic 1. The topic probability of these groups ranges from 0.0 to 1.0, but with a mean closer to a probability of 0.0.

- Topic 2 in this setup was defined as high speed driving. The results depicted in Fig. 4 illustrate a reversed output compared to Topic 1. Here, alert participants have a high probability for Topic 1, and for unalert participants, the mean is closer to 1.0 than 0.0. For unalert participants this means that the probability of highway driving ranges from 0.0 to 1.0.

5   Discussion

The increase of road fatalities has led to the urge to find methods and systems to prevent risky driving behavior. Preferably, these methods need to automated, as human interference might lead to a bias in risky driving detection. Furthermore, current techniques, which record naturalistic driving behavior, have brought the potential and challenge in detection of driving behavior. Large amounts of data being captured automatically have the potential to capture individual driving behavior, which informs us about how people drive. On the other hand, the challenge in analyzing large data sets, is to prevent the loss of important data characteristics, to apply techniques which change the structure of time series data, and to encounter models that are not interpretable, transparent, and explainable. Therefore, this study applied a time-series abstraction method (SAX, Symbolic Aggregate Approximation) together with topic modeling using LDA, which enables explicative approaches making models and results interpretable, transparent and explainable.

In our experiments, we applied symbolic representations (SAX), for which we then applied Latent Dirichlet Allocation (LDA) for probabilistic topic modeling. The topics covered general information about the driving experiments, which were held for this study. The largest topic in our first (simple) model represented city driving, as a state in which the vehicle was stationary was overrepresented. Especially in urban settings, in which roads are more crowded, and traffic lights are more present, it is more likely to stand still with a vehicle. Besides this state, Topic 1 was represented by speeds, which were a reflection of the speed limits that were present in the urban environment. The second topic, represented highway driving, in which SAX-words with high speed occurrences were overrepresented. Furthermore, we analyzed whether more complex topic models would enable more powerful insights for identifying behavioral patterns. Inclusion of n-grams led to new insights in the data, as bi- and trigrams provided more information about occurrences in the data which followed each other. More specifically, besides city- and highway driving, this model could provide more detailed information about the situations that occurred during the experiment. For example, a clear distinction between highway driving, with low density in traffic was revealed, compared to high density (i.e., rush hour). Also, the model detected driving behavior in which a maximum speed of 70 km/h was allowed, but high density of vehicles and traffic lights were present.

In addition, this paper also considered the detection of differences between alert and unalert participants. To address this task, a Psychomotor Vigilance Test was included in the experiments and analysis. The results of the test were used to label participants, and to measure whether there were differences in topic probabilities between alert and unalert participants. In general, topic probabilities were in similar ranges. However, for the more complex topic models, we observed several interesting findings: For example, Topic 1 in the LDA-model including n-grams indicated that four alert drivers had a high topic probability for highway driving, which were the only participants in the sample who drove on a highway.

Compared to the dataset of the study in [27], this study was limited to a sample size of 24 participants, which equalized 24 rides, as a limitation of this current study. Nevertheless, we were already able to identify interesting and characteristic patterns from the collected dataset. A larger sample size would benefit from more nuances in driving data, which would be translated in a potentially broader topic model, with risky driving behavior [17].

6   Conclusions

The results of the current study, have provided different behavioral patterns providing novel insights in LDA-analysis. Furthermore, we have shown that the applied methodology using extended LDA models allows to obtain more comprehensive models for identifying behavioral patterns, which also showed interesting results in the analysis of the alertness of drivers. To the best of our knowledge, this is the first time that such methods have been applied in this context. This contribution might provide the opportunity to detect patterns in naturalistic driving behavior, which would not have been detected with human interference. This way, it is possible to detect pre-accidental situations, which provide more information about the driving behavior of people.

For future work, we aim at analyzing richer behavioral profiles on topic models utilizing subgroup discovery [8,23]. Furthermore, the spatio-temporal analysis of the (abstracted) time-series data using data mining and network analysis [3,7,20,33] as well as contextualized approaches for local exceptionality mining, e.g., [4,9,22] are interesting directions for future research. Extending the analysis approach, and also integrating other datasets is a further aim for a follow-up study.

References

1. Arsintescu, L., Kato, K.H., Cravalho, P.F., Feick, N.H., Stone, L.S., Flynn-Evans, E.E.: Validation of a touchscreen psychomotor vigilance task. Accid. Anal. Prev. **126**, 173–176 (2017)
2. Aryal, A., Ghahramani, A., Becerik-Gerber, B.: Monitoring fatigue in construction workers using physiological measurements. Autom. Constr. **82**, 154–165 (2017)
3. Atzmueller, M.: Data mining on social interaction networks. J. Data Min. Digit. Hum. **1** (2014)
4. Atzmueller, M.: Detecting community patterns capturing exceptional link trails. In: Proceedings of IEEE/ACM ASONAM. IEEE Press, Boston (2016)
5. Atzmueller, M.: Onto explicative data mining: exploratory, interpretable and explainable analysis. In: Proceedings of Dutch-Belgian Database Day. TU Eindhoven (2017)
6. Atzmueller, M.: Declarative aspects in explicative data mining for computational sensemaking. In: Seipel, D., Hanus, M., Abreu, S. (eds.) WFLP/WLP/INAP -2017. LNCS (LNAI), vol. 10997, pp. 97–114. Springer, Cham (2018). https://doi.org/10.1007/978-3-030-00801-7_7
7. Atzmueller, M., Lemmerich, F.: Exploratory pattern mining on social media using geo-references and social tagging information. IJWS **2**(1/2), 80–112 (2013)
8. Atzmueller, M., Puppe, F., Buscher, H.P.: Profiling examiners using intelligent subgroup mining. In: Proceedings of IDAMAP, pp. 46–51, Aberdeen, Scotland (2005)
9. Atzmueller, M., Schmidt, A., Kibanov, M.: DASHTrails: an approach for modeling and analysis of distribution-adapted sequential hypotheses and trails. In: Proceedings of WWW 2016 (Companion). IW3C2/ACM (2016)
10. Basner, M., Dinges, D.F.: Maximizing sensitivity of the psychomotor vigilance test (PVT) to sleep loss. Sleep **34**(5), 581–591 (2011)
11. Bener, A., Lajunen, T., Özkan, T., Yildirim, E., Jadaan, K.S.: The impact of aggressive behaviour, sleeping, and fatigue on road traffic crashes as comparison between minibus/van/pick-up and commercial taxi drivers. Profiling Exam. Using Intell. Subgr. Min. **5**, 21–31 (2017)
12. Blei, D.M.: Probabilistic topic models. CACM **55**(4), 77–84 (2012)
13. Blei, D.M., Ng, A.Y., Jordan, M.I.: Latent Dirichlet allocation. J. Mach. Learn. Res. **3**, 993–1022 (2003)
14. Brodsky, W., Olivieri, D., Chekaluk, E.: Music genre induced driver aggression: a case of media delinquency and risk-promoting popular culture. Music Sci. **1**, 2059204317743118 (2018)
15. Brunnauer, A., Segmiller, F.M., Löschner, S., Grun, V., Padberg, F., Palm, U.: The effects of transcranial direct current stimulation (TDCS) on psychomotor and visual perception functions related to driving skills. Front. Behav. Neurosci. **12**, 16 (2018)

16. Cantin, V., Lavallière, M., Simoneau, M., Teasdale, N.: Mental workload when driving in a simulator: effects of age and driving complexity. Accid. Anal. Prev. **41**(4), 763–771 (2009)
17. Chen, C.P., Zhang, C.Y.: Data-intensive applications, challenges, techniques and technologies: a survey on big data. Inf. Sci. **275**, 314–347 (2014)
18. Chen, H.Y.W., Donmez, B., Hoekstra-Atwood, L., Marulanda, S.: Self-reported engagement in driver distraction: an application of the theory of planned behaviour. Transp. Res. Part F Traffic Psychol. Behav. **38**, 151–163 (2016)
19. Garbarino, S., et al.: Insomnia is associated with road accidents. Further evidence from a study on truck drivers. PLoS one **12**(10), e0187256 (2017)
20. Giannotti, F., Nanni, M., Pinelli, F., Pedreschi, D.: Trajectory pattern mining. In: Proceedings of SIGKDD, pp. 330–339. ACM (2007)
21. Guo, F., Fang, Y.: Individual driver risk assessment using naturalistic driving data. Accid. Anal. Prev. **61**, 3–9 (2013)
22. Harri, J., Filali, F., Bonnet, C.: Mobility models for vehicular Ad Hoc networks: a survey and taxonomy. IEEE Commun. Surv. Tutor. **11**(4), 19–41 (2009)
23. Hendrickson, A.T., Wang, J., Atzmueller, M.: Identifying exceptional descriptions of people using topic modeling and subgroup discovery. In: Ceci, M., Japkowicz, N., Liu, J., Papadopoulos, G.A., Raś, Z.W. (eds.) ISMIS 2018. LNCS (LNAI), vol. 11177, pp. 454–462. Springer, Cham (2018). https://doi.org/10.1007/978-3-030-01851-1_44
24. Jones, M.J., et al.: The psychomotor vigilance test: a comparison of different test durations in elite athletes. J. Sport. Sci. **36**(18), 2033–2037 (2018)
25. Lin, J., Keogh, E., Wei, L., Lonardi, S.: Experiencing SAX: a novel symbolic representation of time series. DMKD **15**(2), 107–144 (2007)
26. Loh, S., Lamond, N., Dorrian, J., Roach, G., Dawson, D.: The validity of psychomotor vigilance tasks of less than 10-minute duration. Behav. Res. Methods Instrum. Comput. **36**(2), 339–346 (2004)
27. McLaurin, E., et al.: Variations on a theme: topic modeling of naturalistic driving data. In: Proceedings of Human Factors and Ergonomics Society Annual Meeting, pp. 2107–2111 (2014)
28. Merino, S., Atzmueller, M.: Behavioral Topic modeling on naturalistic driving data. In: Proceedings of BNAIC. Jheronimus Academy of Data Science, Den Bosch, The Netherlands (2018)
29. Puschmann, D., Barnaghi, P., Tafazolli, R.: Using LDA to uncover the underlying structures and relations in smart city data streams. IEEE Syst. J. **12**(2), 1755–1766 (2018)
30. Saxby, D.J., Matthews, G., Neubauer, C.: The relationship between cell phone use and management of driver fatigue: it's complicated. J. Saf. Res. **61**, 129–140 (2017)
31. Sehgal, S., Kapoor, R.: Mathematical relationship among visual reaction time, age and BMI in healthy adults. Indian J. Appl. Res. **8**(8), 35 (2018)
32. Venkatraman, V., Liang, Y., McLaurin, E.J., Horrey, W.J., Lesch, M.F.: Exploring driver responses to unexpected and expected events using probabilistic topic models. In: Proceedings of International Driving Symposium on Human Factors in Driver Assessment, Training and Vehicle Design, pp. 375–381. University of Iowa (2017)
33. Verhein, F., Chawla, S.: Mining spatio-temporal patterns in object mobility databases. Data Min. Knowl. Discov. **16**(1), 5–38 (2008)
34. Wohleber, R.W., Matthews, G.: Multiple facets of overconfidence: implications for driving safety. Transp. Res. Part F Traffic Psychol. Behav. **43**, 265–278 (2016)

35. Yang, L., Li, X., Guan, W., Zhang, H.M., Fan, L.: Effect of traffic density on drivers' lane change and overtaking manoeuvres in freeway situation: a driving simulator based study. Traffic Inj. Prev. **19**, 1–25 (2018)
36. Zhang, G., Yau, K.K., Zhang, X., Li, Y.: Traffic accidents involving fatigue driving and their extent of casualties. Accid. Anal. Prev. **87**, 34–42 (2016)
37. Zheng, Y., Wang, J., Li, X., Yu, C., Kodaka, K., Li, K.: Driving risk assessment using cluster analysis based on naturalistic driving data. In: Proceedings of International Conference on Intelligent Transportation Systems, pp. 2584–2589. IEEE (2014)

Sequential Monte Carlo Inference Based on Activities for Overlapping Community Models

Shohei Sakamoto[1] and Koji Eguchi[2(✉)]

[1] Kobe University, Kobe 657-8501, Japan
`shohei@cs25.scitec.kobe-u.ac.jp`
[2] Hiroshima University, Higashi-Hiroshima 739-8527, Japan
`eguchi@acm.org`

Abstract. Various kinds of data such as social media can be represented as a network or graph. Latent variable models using Bayesian statistical inference are powerful tools to represent such networks. One such latent variable network model is a Mixed Membership Stochastic Blockmodel (MMSB), which can discover overlapping communities in a network and has high predictive power. Previous inference methods estimate the latent variables and unknown parameters of the MMSB on the basis of the whole observed network. Therefore, dynamic changes in network structure over time are hard to track. Thus, we first present an incremental Gibbs sampler based on node activities that focuses only on observations within a fixed term length for online sequential estimation of the MMSB. We further present a particle filter based on node activities with various term lengths. For instance, in an e-mail communication network, each particle only considers e-mail accounts that sent or received a message within a specific term length, where the length may be different from those of other particles. We show through experiments with two link prediction datasets that our proposed methods achieve both high prediction performance and computational efficiency.

1 Introduction

Many kinds of data can be represented as a network or a graph, which is sometimes dynamic and large in scale. Typical examples of such dynamic, large-scale networks are social networks. By modeling such networks, we can discover communities that have a shared property, so as to avoid high-dimensional difficulties and to visualize complex networks, and can also uncover temporal dynamics in such communities. Moreover, we can predict links or relationships that do exist but have not been observed or do not exist but may appear in the near future. Latent variable models using Bayesian statistical inference are a powerful tool to analyze such networks [7].

In latent variable models for networks, latent random variables are used to represent communities or groups underlying a network, and observed random

© Springer Nature Switzerland AG 2019
M. Atzmueller et al. (Eds.): MUSE 2015/MSM 2015/MSM 2016, LNAI 11406, pp. 89–106, 2019.
https://doi.org/10.1007/978-3-030-34407-8_5

variables are used to represent the nodes. Latent variable models for networks can be classified into two approaches. One is to assume that every node is assigned to a single community [13,17,19]. The other is to assume that every node is assigned to multiple communities, resulting in overlapping communities [1,16]. The latter approach is often more flexible for analyzing real-world complex networks since, for instance, each social actor often belongs to more than one community in social networks. In this paper, we focus on a Mixed Membership Stochastic Blockmodel (MMSB) [1] as a typical latent variable model that provides overlapping communities. In the MMSB, each node is represented by a mixture of latent groups, where each group is represented by a multinomial distribution over nodes. The MMSB is effective for community discovery and link prediction.

The latent variables and unknown parameters of MMSB can be estimated by using variational Bayesian inference [1] or collapsed Gibbs sampling [15]. The MMSB is usually estimated on the basis of a whole observed network. This is called the batch estimation method. However, this method is not suitable in realistic situations, such as when the observations of links are given sequentially. Online estimation methods are promising for addressing these problems; however, previous online estimation methods [15] have room for improvement. The structure of a real-world complex network often changes over time, so old observations of links do not help and can even harm the estimation accuracy. In this paper, we address online estimation problems in such dynamic settings.

We first present an incremental Gibbs sampler based on node activities that focuses only on observations within a fixed term length for online sequential estimation of the MMSB. In an e-mail communication network, the incremental Gibbs sampler only considers e-mail accounts that sent or received a message within a specific term length. We further present a particle filter based on node activities with various term lengths. In an e-mail communication network, each particle only considers e-mail accounts that sent or received a message in a specific term length, where the length may be different from those of other particles. Through experiments with two link prediction datasets: a university community site dataset and an e-mail communication dataset, we show that our proposed incremental Gibbs sampler and particle filter can achieve both high prediction performance and computational efficiency.

The structure of this paper is as follows. In Sect. 2, we briefly describe some related work. We outline MMSB and its estimation methods in Sect. 3 and describe how to take into account the node activities in Sect. 4. We then detail our experiments in Sect. 5, results of which indicated our proposed methods estimate models more effective and efficient than baselines. Finally, we conclude the paper in Sect. 6.

2   Related Work

A number of statistical network models were explored in previous studies, for example, to discover social roles in social network data and predict missing links in biological networks [7]. Latent variable models with Bayesian statistical inference are a powerful tool to analyze such networks [1,13,16,17,19]. Nowicki and

Snijders [17,19] developed a stochastic blockmodel where each node is assigned to a cluster drawn from a multinomial over a fixed, finite number of clusters. Kemp et al. [13] extended stochastic blockmodels to an Infinite Relational Model (IRM) that assumes an infinite number of clusters. Karrer et al. [12] proposed (finite) stochastic blockmodels that take into account variation in vertex degree and achieved more effective performance. These models are based on the assumption that every node is assigned to a single cluster. In contrast, another line of research is based on the assumption that every node is assigned to multiple groups or communities, resulting in overlapping communities [1,16]. The Mixed Membership Stochastic Blockmodel (MMSB) [1] is a typical approach for such overlapping community models, where each node is represented by a mixture of latent groups, and each latent group is represented by a multinomial distribution over nodes. The Latent Feature Relational Model (LFRM) [16] is another example of this line of research. LFRM assumes an infinite number of latent features, assigning each node to more than one latent feature drawn from an Indian Buffet Process [6]. Ball et al. [2] also demonstrated a method for detecting overlapping communities in a different way based on probabilistic latent semantic analysis [10].

The work presented in this paper can be positioned along overlapping community models, especially the MMSB, focusing on sequential (online) inference. The sequential inference has two advantages. First, it can help the problem that the inference for the overlapping community models is usually expensive. Second, it can tracks dynamical changes in network structure over time. As for the previous work on sequential inference, Iwata et al. [11] considered multiple timescales for analyzing documents with time stamps. Since the objective of their study was text analysis, it is essentially different from our work on network analysis. Gopalan et al. [8] developed an efficient estimation for overlapping community models, where only random samples of observed links are used for the inference. They focused on efficiency in estimating the latent variables and unknown parameters of overlapping community models, not temporal dynamics in network structure. Kobayashi and Eguchi [15] explored sequential inference with overlapping community models in online settings. However, they did not consider dynamic phenomena when the structure of a real-world complex network often changes over time and therefore old observations of links do not help and can even harm the inference. This paper presents sequential inference based on an incremental Gibbs sampler and a particle filter that uses node activities, which has not been explored in previous studies on temporal dynamics in modeling overlapping communities.

3 Mixed Membership Stochastic Blockmodel

A Mixed Membership Stochastic Blockmodel (MMSB) [1] is an overlapping community model for network data. In this section, we first outline the modeling of the MMSB and then review the inference methods in both batch and online settings [15].

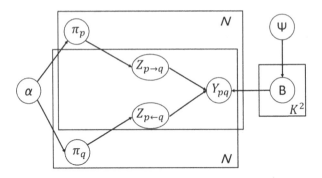

Fig. 1. Graphical model of MMSB.

3.1 Modeling

First, we give the definitions used in this paper. We represent a graph as $\mathbf{G} = (\mathbf{N}, \mathbf{Y})^1$, where $\mathbf{N}$ is a set of nodes or vertices, and (p, q) element in adjacency matrix $\mathbf{Y}$ indicates whether a link or arc is absent or present from node p to node q as $Y(p, q) \in \{0, 1\}$. Each node is associated with a multinomial $\mathbf{Mult}(\boldsymbol{\pi}_p)$ over latent groups or communities (hereinafter, just "groups"), assuming a Dirichlet prior $\mathbf{Dir}(\boldsymbol{\alpha})$ over multinomial parameters $\boldsymbol{\pi}_p = \{\pi_{p,g} : g \in \{1, \cdots, K\}\}$. Here, $\pi_{p,g}$ indicates node p's multinomial parameter for any group $g \in \{1, \cdots, K\}$, representing the probability that node p falls into group g. Relationships between each pair of groups are defined by matrix $B_{K \times K}$ where each element represents a Bernoulli parameter with a Beta prior $\mathbf{Beta}(\boldsymbol{\psi})$. Here, $B(g, h)$ indicates the probability of generating a link from an arbitrary node in group g to another arbitrary node in group h. Given a link from p to q, indicator vector $\mathbf{z}_{p \rightarrow q}$ represents a group assigned to p, and $\mathbf{z}_{p \leftarrow q}$ represents a group assigned to q. These indicator vectors are denoted by $\mathbf{Z}_{\rightarrow} = \{\mathbf{z}_{p \rightarrow q} : p, q \in \mathbf{N}\}$ and $\mathbf{Z}_{\leftarrow} = \{\mathbf{z}_{p \leftarrow q} : p, q \in \mathbf{N}\}$. In accordance with the definitions above, the generative process of MMSB can be described as follows.

1. For each node p:
   - Draw a K-dimensional vector of multinomial parameters, $\boldsymbol{\pi}_p \sim \mathbf{Dir}(\boldsymbol{\alpha})$
2. For each pair of groups (g, h):
   - Draw a Bernoulli parameter, $B(g, h) \sim \mathbf{Beta}(\boldsymbol{\psi}(g, h))$
3. For each pair of nodes (p, q):
   - Draw an indicator vector for the initiator's group assignment, $\mathbf{z}_{p \rightarrow q} \sim \mathbf{Mult}(\boldsymbol{\pi}_p)$
   - Draw an indicator vector for the receiver's group assignment, $\mathbf{z}_{p \leftarrow q} \sim \mathbf{Mult}(\boldsymbol{\pi}_q)$
   - Sample a binary value that represents the presence or absence of a link, $Y(p, q) \sim \mathbf{Bern}(\mathbf{z}_{p \rightarrow q}^{\mathrm{T}} \mathbf{B} \mathbf{z}_{p \leftarrow q})$

---

[1] In this paper, we assume a directed graph for representing the network structure, but the network structure can also be easily applied to an undirected graph.

Algorithm 1: batch Gibbs sampler for $N \times N$
1: initialize group assignment randomly for $N \times N$
2: for iter=1 to L_{iter} do
3:   for p=1 to N do
4:     for q=1 to N do
5:       draw $z_{p \to q}$ from $P(\mathbf{z}_{p \to q} \mid \mathbf{Y}, \mathbf{Z}_{\to}^{\neg(p,q)}, \mathbf{Z}_{\leftarrow}^{\neg(p,q)}, \alpha, \Psi)$
6:       draw $z_{p \leftarrow q}$ from $P(\mathbf{z}_{p \leftarrow q} \mid \mathbf{Y}, \mathbf{Z}_{\to}^{\neg(p,q)}, \mathbf{Z}_{\leftarrow}^{\neg(p,q)}, \alpha, \Psi)$
7:     end for
8:   end for
9: end for
10: complete the posterior estimates of π and $\mathbf{B}$

The joint distribution with all the random variables (full joint distribution) is given as follows:

$$P(\mathbf{Y}, \pi_{1:N}, \mathbf{Z}_{\to}, \mathbf{Z}_{\leftarrow}, \mathbf{B} \mid \alpha, \Psi)$$
$$= P(\mathbf{B} \mid \Psi) \prod_{p,q:p \neq q} P(Y(p,q) \mid \mathbf{z}_{p \to q}, \mathbf{z}_{p \leftarrow q}, \mathbf{B}) P(\mathbf{z}_{p \to q} \mid \pi_p)$$
$$P(\mathbf{z}_{p \leftarrow q} \mid \pi_q) \prod_p P(\pi_p \mid \alpha) \tag{1}$$

A graphical model representation of the MMSB is shown in Fig. 1.

3.2 Batch Gibbs Sampler

Next, we describe a batch Gibbs sampler for estimating the latent variables and unknown parameters of an MMSB. For an observed link from node p to node q, the full conditional probability of assigning groups g and h to p and q, respectively, is given by:

$$P(z_{p \to q} = g, z_{p \leftarrow q} = h \mid \mathbf{Y}, \mathbf{Z}_{\to}^{\neg(p,q)}, \mathbf{Z}_{\leftarrow}^{\neg(p,q)}, \alpha, \psi)$$
$$\propto (n(p,g) - 1 + \Delta(g' \neq g) + \alpha_g)(n(q,h) - 1 + \Delta(h' \neq h) + \alpha_h)$$
$$\frac{n(g,h,\delta) - 1 + \Delta(g' \neq g \wedge h' \neq h) + \psi_\delta}{n(g,h,0) + n(g,h,1) - 1 + \Delta(g' \neq g \wedge h' \neq h) + \psi_0 + \psi_1}$$
$$= \begin{cases} \pi_{p,g}^{\neg(p,q)} \pi_{q,h}^{\neg(p,q)} B(g,h)^{\neg(p,q)} & (\text{if } \delta = 1) \\ \pi_{p,g}^{\neg(p,q)} \pi_{q,h}^{\neg(p,q)} (1 - B(g,h)^{\neg(p,q)}) & (\text{if } \delta = 0) \end{cases} \tag{2}$$

where $z_{p \to q}$ and $z_{p \leftarrow q}$ are the latent random variable[2] representing group assignments to initiator node p and receiver node q, respectively, as mentioned in the

---

[2] In Sect. 3.1, we used an indicator vector $\mathbf{z}_{p \to q}$ where a specific component is one, corresponding to the group indicated by $z_{p \to q}$, and all the others are zero, for convenience.

previous section. $n(p, g)$ indicates the count of g assigned to p. $n(g, h, \delta)(\delta \in \{0, 1\})$ indicates the count of presence $(\delta = 1)$ or absence $(\delta = 0)$ of links, where any initiator node is assigned to g and any receiver node is assigned to h. Moreover, α_g indicates g-th component of K-dimensional vector of Dirichlet hyperparameter α. ψ_1 and ψ_0 indicate Beta hyperparameters corresponding to the presence and absence of links, respectively. "$\neg(p, q)$" indicates ignoring the current group assignment to the link from p to q. The indicator function $\Delta(\cdot)$ takes one when the designated event occurs and zero if otherwise. g' and h' indicate the groups currently assigned to p and q, respectively.

In general, many kinds of real-world networks are sparse, as can often be seen in social networks. Therefore, there are often zero elements in the adjacency matrix $\mathbf{Y}$. To avoid bias due to this nature of networks, a sparsity parameter ρ is sometimes introduced into Eq. (2), as in the work of [1]:

$$P(z_{p \to q} = g, z_{p \leftarrow q} = h | \mathbf{Y}, \mathbf{Z}_{\to}^{\neg(p,q)}, \mathbf{Z}_{\leftarrow}^{\neg(p,q)}, \alpha, \psi)$$

$$\propto \begin{cases} (1 - \rho)\pi_{p,g}^{\neg(p,q)}\pi_{q,h}^{\neg(p,q)}B(g, h)^{\neg(p,q)} & (\text{if } \delta = 1) \\ (1 - \rho)\pi_{p,g}^{\neg(p,q)}\pi_{q,h}^{\neg(p,q)}(1 - B(g, h)^{\neg(p,q)}) + \rho \ (\text{if } \delta = 0) \end{cases} \quad (3)$$

where ρ is given by:

$$\rho = 1 - \sum_{p,q} \frac{Y(p, q)}{N(N - 1)} \quad (4)$$

By using the full conditional probability in Eq. (3), a collapsed Gibbs sampler [9,15] estimates latent variables and unknown parameters of MMSB.

We outline the estimation procedure in Algorithm 1, in which a posterior can be obtained when it converges within L_{iter} iterations. This estimation algorithm is called a batch Gibbs sampler.

3.3 Incremental Gibbs Sampler

The batch Gibbs sampler described in Sect. 3.2 uses the whole observed network, so it does not suit an online situation where new nodes and links are observed sequentially. An incremental Gibbs sampler [4] works in such online situations. It was used to estimate Latent Dirichlet Allocation [3] for text data and was also applied to an MMSB for network data [15].

Given a time series of network data, the batch Gibbs sampler is used for the first certain period of the data. Then, we go through the following steps every time the presence or absence of a link is observed. More details are shown in Algorithm 2.

1. When a new link with an existing node is observed, we sample a pair of groups in accordance with the full conditional probability with already observed data and their group assignments on the basis of Eq. (3), as shown in lines 3 and 4 in Algorithm 2.
2. When a new link with a new node is observed, we sample a pair of groups for every pair of a new node and an already observed node, as shown in lines 5 to 16 in Algorithm 2.

Algorithm 2: incremental Gibbs sampler

1: batch Gibbs sampler for $N_{firstTerm} \times N_{firstTerm}$
2: while(add link $p' \rightarrow q'$) do
3:　　draw $\mathbf{z}_{p' \rightarrow q'}$ from $P(\mathbf{z}_{p' \rightarrow q'} | \mathbf{Y}, \mathbf{Z}_{\rightarrow}^{\neg(p',q')}, \mathbf{Z}_{\leftarrow}^{\neg(p',q')}, \alpha, \psi)$
4:　　draw $\mathbf{z}_{p' \leftarrow q'}$ from $P(\mathbf{z}_{p' \leftarrow q'} | \mathbf{Y}, \mathbf{Z}_{\rightarrow}^{\neg(p',q')}, \mathbf{Z}_{\leftarrow}^{\neg(p',q')}, \alpha, \psi)$
5:　　if(p' is new node) then
6:　　　　for $q = 1$ to $N_{current}$ (if $q \neq q'$) do
7:　　　　draw $\mathbf{z}_{p' \rightarrow q}$ from $P(\mathbf{z}_{p' \rightarrow q} | \mathbf{Y}, \mathbf{Z}_{\rightarrow}^{\neg(p',q)}, \mathbf{Z}_{\leftarrow}^{\neg(p',q)}, \alpha, \psi)$
8:　　　　draw $\mathbf{z}_{p' \leftarrow q}$ from $P(\mathbf{z}_{p' \leftarrow q} | \mathbf{Y}, \mathbf{Z}_{\rightarrow}^{\neg(p',q)}, \mathbf{Z}_{\leftarrow}^{\neg(p',q)}, \alpha, \psi)$
9:　　　　end for
10:　end if
11:　if(q' is new node) then
12:　　　　for $p = 1$ to $N_{current}$ (if $p \neq p'$) do
13:　　　　draw $\mathbf{z}_{p \rightarrow q'}$ from $P(\mathbf{z}_{p \rightarrow q'} | \mathbf{Y}, \mathbf{Z}_{\rightarrow}^{\neg(p,q')}, \mathbf{Z}_{\leftarrow}^{\neg(p,q')}, \alpha, \psi)$
14:　　　　draw $\mathbf{z}_{p \leftarrow q'}$ from $P(\mathbf{z}_{p \leftarrow q'} | \mathbf{Y}, \mathbf{Z}_{\rightarrow}^{\neg(p,q')}, \mathbf{Z}_{\leftarrow}^{\neg(p,q')}, \alpha, \psi)$
15:　　　　end for
16:　end if
17:　for($p'' \rightarrow q''$ in $\mathcal{R}(p',q')$)
18:　　　draw $\mathbf{z}_{p'' \rightarrow q''}$ from $P(\mathbf{z}_{p'' \rightarrow q''} | \mathbf{Y}, \mathbf{Z}_{\rightarrow}^{\neg(p'',q'')}, \mathbf{Z}_{\leftarrow}^{\neg(p'',q'')}, \alpha, \psi)$
19:　　　draw $\mathbf{z}_{p'' \leftarrow q''}$ from $P(\mathbf{z}_{p'' \leftarrow q''} | \mathbf{Y}, \mathbf{Z}_{\rightarrow}^{\neg(p'',q'')}, \mathbf{Z}_{\leftarrow}^{\neg(p'',q'')}, \alpha, \psi)$
20:　end for
21: end while
22: complete the posterior estimates of π and $\mathbf{B}$

---

3. We update the latent groups for the *rejuvenation sequence* $\mathcal{R}(p,q)$—i.e., $|\mathcal{R}(p,q)|$ of randomly selected node pairs that had already been observed when node pair (p,q) was observed, as shown in lines 17 to 20 in Algorithm 2. This step is called *rejuvenation* in the literature on particle filters [5].

We should skip the rejuvenation step or set $|\mathcal{R}(p,q)| = 0$ when network structure frequently changes over time. The incremental Gibbs sampler is extended to the particle filter described below.

3.4   Particle Filter

Particle filters, also known as sequential Monte Carlo methods, are based on the weighted average of multiple particles [4,5,15]. Each particle estimates group assignments for observed node pairs differently from the other particles in accordance with the steps of the incremental Gibbs sampler described in Sect. 3.3. The weight of each particle represents its importance, which is updated by using the likelihood of generating the observed link. When the variance of the weight is larger than a predefined threshold referred to as effective sample size (ESS) threshold, resampling is performed to make a new set of particles where the particles with negligibly low weights are replaced by new particles copied from those with higher weights. The simplest resampling scheme draws particles from the multinomial specified by the normalized weights [5]. After the resampling, the weights are reset to P^{-1}, where P indicates the number of particles. The algorithm is outlined in Algorithm 3.

Algorithm 3: particle filter

1: initialize weights $\omega^{(k)} = P^{-1}$ for $k = 1, \cdots, P$

2: while(add link $p' \to q'$)

3:    for $k = 1$ to P do

4:      $\omega^{(k)} \mathrel{*}= P^{(k)}(Y(p',q') = 1|\mathbf{Y}, \mathbf{Z}_{\to}^{\neg(p',q')}, \mathbf{Z}_{\leftarrow}^{\neg(p',q')}, \boldsymbol{\alpha}, \boldsymbol{\psi})$

5:      draw $z_{p' \to q'}^{(k)}$ from $P^{(k)}(\mathbf{z}_{p' \to q'}|\mathbf{Y}, \mathbf{Z}_{\to}^{\neg(p',q')}, \mathbf{Z}_{\leftarrow}^{\neg(p',q')}, \boldsymbol{\alpha}, \boldsymbol{\psi})$

6:      draw $z_{p' \leftarrow q'}^{(k)}$ from $P^{(k)}(\mathbf{z}_{p' \leftarrow q'}|\mathbf{Y}, \mathbf{Z}_{\to}^{\neg(p',q')}, \mathbf{Z}_{\leftarrow}^{\neg(p',q')}, \boldsymbol{\alpha}, \boldsymbol{\psi})$

7:      if(p' is new node)

8:        for $q = 1$ to $N_{current}$ (if $q \neq q'$)

9:          draw $z_{p' \to q}^{(k)}$ from $P^{(k)}(\mathbf{z}_{p' \to q}|\mathbf{Y}, \mathbf{Z}_{\to}^{\neg(p',q)}, \mathbf{Z}_{\leftarrow}^{\neg(p',q)}, \boldsymbol{\alpha}, \boldsymbol{\psi})$

10:         draw $z_{p' \leftarrow q}^{(k)}$ from $P^{(k)}(\mathbf{z}_{p' \leftarrow q}|\mathbf{Y}, \mathbf{Z}_{\to}^{\neg(p',q)}, \mathbf{Z}_{\leftarrow}^{\neg(p',q)}, \boldsymbol{\alpha}, \boldsymbol{\psi})$

11:        end for

12:      end if

13:      if(q' is new node)

14:        for $p = 1$ to $N_{current}$ (if $p \neq p'$) do

15:          draw $z_{p \to q'}^{(k)}$ from $P^{(k)}(\mathbf{z}_{p \to q'}|\mathbf{Y}, \mathbf{Z}_{\to}^{\neg(p,q')}, \mathbf{Z}_{\leftarrow}^{\neg(p,q')}, \boldsymbol{\alpha}, \boldsymbol{\psi})$

16:         draw $z_{p \leftarrow q'}^{(k)}$ from $P^{(k)}(\mathbf{z}_{p \leftarrow q'}|\mathbf{Y}, \mathbf{Z}_{\to}^{\neg(p,q')}, \mathbf{Z}_{\leftarrow}^{\neg(p,q')}, \boldsymbol{\alpha}, \boldsymbol{\psi})$

17:        end for

18:      end if

19:    end for

20:    normalize weights ω to sum to 1

21:    if $|\omega|^{-2} \leq$ ESS threshhold then

22:      resample particles

23:      $\omega^{(k)} = P^{-1}$ for $k = 1, \cdots, P$

24:    end if

25:    for $k = 1$ to P do

26:      for($p'' \to q''$ in $\mathcal{R}(p',q')$) do

27:        draw $z_{p'' \to q''}^{(k)}$ from

28:          $P^{(k)}(\mathbf{z}_{p'' \to q''}|\mathbf{Y}, \mathbf{Z}_{\to}^{\neg(p'',q'')}, \mathbf{Z}_{\leftarrow}^{\neg(p'',q'')}, \boldsymbol{\alpha}, \boldsymbol{\psi})$

29:        draw $z_{p'' \leftarrow q''}^{(k)}$ from

30:          $P^{(k)}(\mathbf{z}_{p'' \leftarrow q''}|\mathbf{Y}, \mathbf{Z}_{\to}^{\neg(p'',q'')}, \mathbf{Z}_{\leftarrow}^{\neg(p'',q'')}, \boldsymbol{\alpha}, \boldsymbol{\psi})$

31:      end for

32:    end for

33: end while

34: complete the posterior estimates of $\boldsymbol{\pi}$ and $\mathbf{B}$

The particle filter gives a posterior as:

$$P_{particle} = \sum_k (P^{(k)} \times \omega^{(k)}) \tag{5}$$

where $P^{(k)}$ indicates the posterior given by k-th particle in accordance with Eq. (3). $\omega^{(k)}$ indicates the weight of the k-th particle, which is proportional to the likelihood of generating observed links by using the particle. The incremental Gibbs sampler in Sect. 3.3 can be considered as a special case when the number of particles is one.

4   Online Inference Using Node Activities

The structure of a real-world complex network often changes over time, so old observations of links do not help and can even harm the estimation accuracy. The previous online estimation methods in Sect. 3 have some drawbacks in such dynamic settings. When a link from/to an unseen node is observed, the previous online estimation methods assume the absence of links between the unseen node and the other nodes that were observed previously, except for the node that is currently linked to/from the unseen node. Thus, they assign groups to not only the pair of nodes that form the observed link but also all the other nodes that were observed previously. However, groups are sometimes inappropriately assigned on the basis of this assumption since some of the node pairs *potentially* have links that have no chance to be observed. Suppose an example of friendships in social media, we have two types of observations when a newcomer ('X') becomes a friend with someone else ('Y'); one is the observation that 'X' is a friend of 'Y', and the other is that 'X' is non-friends of all the other people except for 'Y'. The non-friend observations are sometimes wrong, since some of them may be missing observations –they are actually friends but not known to the other people–. In the following section, we take into account *node activities* to address this problem by assuming that the non-friend observations are only applied to *active people* who become a friend of someone a short time ago.

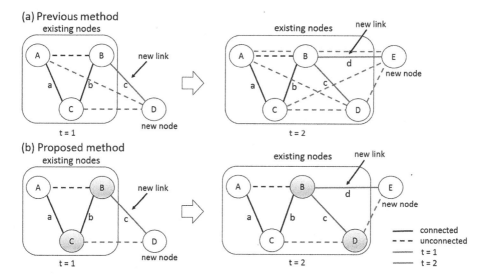

Fig. 2. Illustration of how to estimate group assignments using previous methods (top) and proposed methods (bottom) when $\ell = 1$. Active nodes are shaded in this figure.

4.1   Definition of Node Activities

We first define *active* nodes as those that were observed to be linked to/from others during a period of observing the most recent ℓ links, while all the other already observed nodes are defined as *inactive*. We then assign groups on the assumption that, when a link is observed from/to an unseen node, the absence of links is also observed only to/from the active nodes. Figure 2 illustrates the flow of estimating the group assignments by using the previous and proposed methods. In this figure, $a, b, \cdots, d$ represents the links that are observed in alphabetical order. When the presence of a link is observed between a new node D and an already observed node B at time $t = 1$, the previous methods assign groups to the pair of nodes (D, B) and also to (D, A) and (D, C), assuming that A and C had already been observed by that time and that the absence of links is observed for (D, A) and (D, C) at that time. On the other hand, in the same situation, the proposed incremental Gibbs sampler and particle filter assign groups to the pair of nodes (D, B) and (D, C) but not (D, A) when C is active (observed to have a link closely by that time) but A is inactive (not observed to have had a link for a while). When $t = 2$, groups are assigned on the same assumption. In this way, the proposed methods assign groups only to active nodes when a new node is observed, while the previous methods assign groups to all the already observed nodes in the same situation, regardless of whether the new node is linked to/from the already observed nodes. Thus, the proposed methods may work to avoid inappropriate group assignments that are caused by the frequent observations of absence of links, making model estimation more accurate. In addition, since estimation of group assignments for non-active nodes (treated as missing values) are not performed, the computational cost can be expected to be reduced.

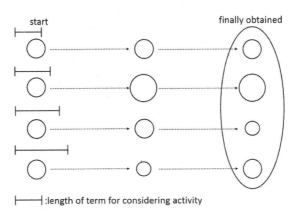

Fig. 3. Illustration of particle filter using node activities in various term lengths.

4.2 Incremental Gibbs Sampler and Particle Filter Using Node Activities

By simply applying the node activities discussed in the previous section, we can modify the particle filter described in Sect. 3.4. We can also modify the incremental Gibbs sampler in Sect. 3.3 by assuming that the number of particles in the particle filter is one.

The node activities rely on a term length since the node is deemed to be active if some activities are observed within the term but inactive if otherwise. From this consideration, we propose using a particle filter based on node activities with various term lengths by setting a different term length for each particle, as illustrated in Fig. 3. To calculate the final posterior for this particle filter, we can use Eq. (5) in the same way of the particle filter that is previously described in Sect. 3.4. Note that each particle is now assumed to consider a different term length to detect the node activities.

For instance, in an e-mail communication network, each particle only considers e-mail accounts that sent or received a message within a specific term length, where the length may be different from those of other particles. This particle filter inference may be made more robust by considering multiple terms for the node activities instead of uniformly setting a fixed term length for every particle. We believe that considering multiple terms for the node activities is especially effective to track dynamical changes in network structure over time. The node activities based on a long term should work effectively when the network structure is stable over time. When the network structure changes drastically, the long-term node activities cause inappropriate estimation, so the term should be shorter. Assuming a realistic situation where we do not know the network dynamics in advance, this kind of diversity of particles should work effectively for sequential (online) estimation. We assume that a term length for each particle is sampled from a Poisson distribution, as below:

$$\ell \sim Poisson(\lambda)$$

where λ is the Poisson mean parameter. A Gaussian can also be used instead, but a Poisson is more appropriate since it generates positive integers.

5 Experiments

In this section, we evaluate our methods for the online sequential estimation of a MMSB through experiments with time-series network data. We then discuss both the prediction performance and computational efficiency in time.

5.1 Settings

Datasets. We used two online communication datasets in our experiments, since our ideas on node activities are expected to work more effectively with such dynamic networks.

Dataset A. This dataset is extracted from an online community site for students at the University of California, Irvine from April to October 2004 [18]. A link is assumed to occur when a message is sent from one user to another. The number of nodes is 1,899, and the number of links at each half-moon interval is shown in Fig. 4(a).

Dataset B. This dataset is extracted from the Enron e-mail communication archive [14] from December 1999 to March 2002 and further cleaned so that only the users (i.e., e-mail addresses) who sent and received at least seven e-mails are included [15]. Each node represents an e-mail address, and each link represents an e-mail communication from a sender to a receiver with a time stamp. The number of nodes is 2,356, and the number of links at each time interval is shown in Fig. 4(b).

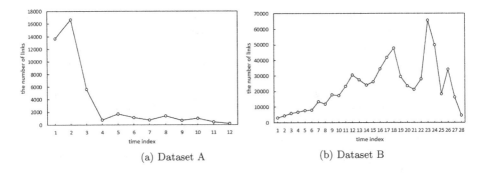

(a) Dataset A   (b) Dataset B

Fig. 4. Number of links at each time interval in Datasets A and B.

Cross-Validation Settings. To use five-fold cross-validation, we divided a set of observations (i.e., each link of which is observed to be present or absent for a pair of nodes) in each dataset evenly into five sets, keeping the time order of the observations. We further divided each set into a test set and a validation set and used the remaining four sets as training sets. Since we estimated a model in an online setting, we used the observations sequentially within the training set. We determined the number of groups and hyperparameters by a grid search using the training and validation sets. We then evaluated the estimation methods using the test set.

Number of Groups and Hyperparameters. Before detailing the online experiments, we describe how we set the number of groups K and hyperparameters α, ψ_0 and ψ_1. We determined the number of groups K by using a grid search over $\{5, 10, 15, \cdots, 40\}$ in a batch setting, resulting in $K = 10$ for Dataset A and $K = 30$ for Dataset B. We also determined ψ_0 and ψ_1 in the manner above, as shown in the results in Tables 1 and 2. For α, we used the symmetric Dirichlet

Table 1. Estimated hyperparameters for Dataset A.

set	ψ_0	ψ_1
set1	0.3456	0.03321
set2	0.3504	0.03053
set3	0.3341	0.02973
set4	0.3110	0.02274
set5	0.3421	0.02839

Table 2. Estimated hyperparameters for Dataset B.

set	ψ_0	ψ_1
set1	0.2411	0.01028
set2	0.2449	0.00864
set3	0.2348	0.00774
set4	0.2569	0.00886
set5	0.2388	0.00915

hyperparameter fixed at 0.1. For a fair comparison, these settings are determined in the same manner used by [15].

When we assumed a fixed term length for the node activities, we determined the term $\ell = 60$ for Dataset A and $\ell = 30$ for Dataset B by a grid search using training and validation sets for each dataset, as shown in Fig. 5. Note that the previous online estimation methods [15] correspond to the case when term length parameter ℓ is large enough. Therefore, the ideas of the node activities contribute to the improvement, as can be seen in this figure. When we assumed various term lengths sampled from a Poisson as discussed in Sect. 4.2, the results of the grid search using training and validation sets were shown in Fig. 6. From this figure, we determined the Poisson mean parameter to be $\lambda = 70$. In these experiments with particle filters, we determined the ESS threshold by doing a grid search over $\{4, 8, 12, 16, 20\}$, fixing the number of particles to 24.

Inference Methods. For estimating MMSB in online settings, we compare the proposed incremental Gibbs sampler using the node activities and the previous incremental Gibbs sampler. Also, we compare two versions of the proposed particle filter using the node activities and the previous particle filter. In our experiments, we set $|\mathcal{R}(p, q)| = 0$ to skip the rejuvenation step in particle filters since it is inappropriate when network structure frequently changes over time.

Evaluation Metrics. We evaluate the prediction performance using the average value of the rate of change of the test-set log-likelihood. The likelihood of

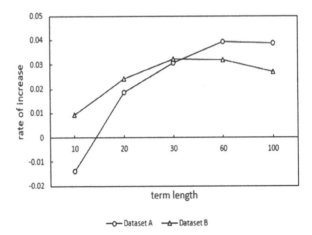

Fig. 5. Results of grid search for term length parameter for Datasets A and B.

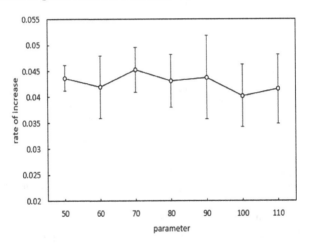

Fig. 6. Sensitivity of Poisson mean parameter in proposed particle filter with various term lengths, in terms of average increase rate of validation-set log-likelihood. Error bars represent sample standard deviations.

the test set $\mathbf{s}^{test}$ indicates how effectively the model predicts unseen data at time interval t using the model estimated with observed data by $t-1$ and is given by:

$$p(\mathbf{s}_{test}^{(t)}) =$$

$$\prod_{(p,q) \in \mathbf{s}_{test}^{(t)}} \sum_{g,h} \left[(1 - \rho^{(t-1)}) \pi_{p,g}^{(t-1)} \pi_{q,h}^{(t-1)} B(g,h)^{(t-1)} \right]^{\delta(p,q)}$$

$$\left[(1 - \rho^{(t-1)}) \pi_{p,g}^{(t-1)} \pi_{q,h}^{(t-1)} (1 - B(g,h)^{(t-1)}) + \rho^{(t-1)} \right]^{1-\delta(p,q)}$$

$$(6)$$

where $\delta(p,q) \in \{1,0\}$ represents the presence or absence of a link from node p to node q. ρ is the sparsity parameter, as defined in Eq. (4). Multinomial parameters $\pi_{p,g}$ and $\pi_{q,h}$ and Bernoulli parameter $\mathbf{B}(g,h)$ are estimated using Eq. (3) using the observations by time $t-1$. Given a discrete-time series network at $t \in \{1, ..., T\}$, the average increase rate of test-set log-likelihood is:

$$\frac{1}{T} \sum_{t=1}^{T} \frac{X(t) - I_0(t)}{|I_0(t)|} \qquad (7)$$

where $X(t)$ represents test-set log-likelihood that has the target inference algorithm at time interval t. $I_0(t)$ represents test-set log-likelihood that is the baseline at t. Here, the baseline is set to the previous incremental Gibbs sampler at t in our experiments.

Since the density of the network is different for each time interval, we evaluated the proposed methods by using the average increase rate of test-set log-likelihood as shown above, not the average test-set log-likelihood itself, which is highly dependent on the number of observations at each time interval. The larger the average increase rate of test-set log-likelihood, the better the target method's prediction performance compared with the baseline.

5.2   Results

In this section, we show the experimental results using the datasets described previously in terms of prediction performance and computational efficiency in time.

Results of Incremental Gibbs Sampler Using Node Activities. Table 3 shows the experimental results of the proposed incremental Gibbs sampling for Datasets A and B, in terms of the average increase rate of test-set log-likelihood, as defined in Sect. 5.1. As mentioned previously, the baseline is the previous method described in Sect. 3.3. From Table 3, the proposed incremental Gibbs sampler with node activities outperformed that without node activities in both cases with two datasets. This indicates that using node activities makes the online sequential inference more flexible.

We next evaluate the time required for the model estimation. Figure 7 demonstrates the time for estimation with each method in time-series plots for Datasets

Table 3. Average increase rate of test-set log-likelihood and its sample standard deviation for incremental Gibbs samplers with Datasets A and B.

Dataset	Dataset A	Dataset B
Activity-based	0.0369 ± 0.0049	0.0322 ± 0.0087
Non-activity-based (Baseline)	(0.0000 ± 0.0000)	(0.0000 ± 0.0000)

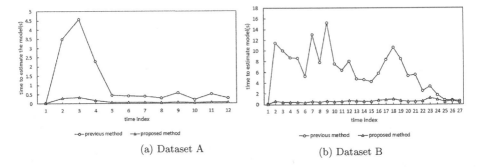

(a) Dataset A   (b) Dataset B

Fig. 7. Evaluation results on the time for estimation in time series plots for Datasets A and B.

A and B. As shown in these graphs, the proposed incremental Gibbs sampler needed less time for estimation than the previous one. This indicates that the proposed method improves computational efficiency in time by considering the node activities. Furthermore, the proposed method brought about a larger improvement for Dataset B than for Dataset A. This is because the proposed method works more efficiently when the number of observed links is larger and so the number of newly observed nodes is also larger, as in Dataset B. Figure 4 shows the number of links in the datasets. This indicates that the proposed method can reduce the time for model estimation for a larger-scale network.

Results of Particle Filter Using Node Activities. Next, we evaluated the particle filter based on the node activities with a fixed term length and that with various term lengths. Table 4 shows the prediction performance results of the proposed particle filters for Datasets A and B. In this table, 'Proposed (Poisson)' indicates the particle filter described in Sect. 4.2 and 'Proposed (fixed-term)' indicates the particle filter using the fixed-term node activities. The baseline is the previous method described in Sect. 3.3, as mentioned previously. As can be seen in Tables 3 and 4, the node activities are helpful for the particle filter inference. Moreover, the particle filter using the node activities with various term lengths works more effectively than that using the fixed-term node activities. From these results, such diverse particles achieve robust sequential (online) estimation, especially for dynamic networks.

Table 4. Average increase rate of test-set log-likelihood and its sample standard deviation for particle filters with Datasets A and B.

Dataset	Dataset A	Dataset B
Activity-based (fixed-term)	0.0373 ± 0.0046	0.0321 ± 0.0055
Activity-based (Poisson)	0.0420 ± 0.0043	0.0326 ± 0.0037
Non-activity-based (Baseline)	(0.0000 ± 0.0000)	(0.0000 ± 0.0000)

From the two evaluations above, the proposed methods are better than the previous methods in terms of both the prediction performance and the estimation time.

6 Conclusions

In this paper, we proposed using an incremental Gibbs sampler and a particle filter for an online sequential estimation of MMSB by considering node activities to track dynamical changes in network structure over time. Our proposed incremental Gibbs sampler and particle filter can reduce the estimation time by using only *active* subsamples of observations in a network. We further explored more flexible estimation by assuming various term lengths for each particle in the particle filter inference.

We used two datasets in our experiments and evaluated the proposed methods in terms of prediction performance and computational efficiency. In terms of prediction performance, the proposed methods were better than the previous methods. In terms of computational efficiency, the proposed methods reduced the cost in estimation time compared with the baselines.

Evaluation under more practical situations are left for future work. Another direction for future work is to apply our ideas to various types of statistical network models. We are especially interested in online sequential estimation of nonparametric relational models, such as a Latent Feature Relational Model (LFRM) [16].

Acknowledgments. This work was supported in part by the Grant-in-Aid for Scientific Research (#15H02703) from JSPS, Japan.

References

1. Airoldi, E.M., Blei, D.M., Fienberg, S.E., Xing, E.P.: Mixed membership stochastic blockmodels. J. Mach. Learn. Res. **9**, 1981–2014 (2008)
2. Ball, B., Karrer, B., Newman, M.E.J.: Efficient and principled method for detecting communities in networks. Phys. Rev. E **84**(3), 036103 (2011)
3. Blei, D.M., Ng, A.Y., Jordan, M.I.: Latent Dirichlet allocation. J. Mach. Learn. Res. **3**, 993–1022 (2003)

4. Canini, K.R., Shi, L., Griffiths, T.L.: Online inference of topics with latent Dirichlet allocation. In: Proceedings of the 12th International Conference on Artificial Intelligence and Statistics, Clearwater Beach, Florida, USA, pp. 65–72 (2009)

5. Doucet, A., de Freitas, N., Gordon, N. (eds.): Sequential Monte Carlo Methods in Practice. Springer, New York (2001). https://doi.org/10.1007/978-1-4757-3437-9

6. Ghahramani, Z., Griffiths, T.L.: Infinite latent feature models and the Indian buffet process. In: Advances in Neural Information Processing Systems, vol. 18 (2006)

7. Goldenberg, A., Zheng, A.X., Fienberg, S.E., Airoldi, E.M.: A survey of statistical network models. Found. Trends Mach. Learn. **2**(2), 129–233 (2010)

8. Gopalan, P.K., Gerrish, S., Freedman, M., Blei, D.M., Mimno, D.M.: Scalable inference of overlapping communities. In: Advances in Neural Information Processing Systems, vol. 25 (2012)

9. Griffiths, T.L., Steyvers, M.: Finding scientific topics. Proc. Natl. Acad. Sci. U. S. A. **101**, 5228–5235 (2004)

10. Hofmann, T.: Probabilistic latent semantic indexing. In: Proceedings of the 22nd Annual International ACM SIGIR Conference on Research and Development in Information Retrieval, Berkeley, California, USA, pp. 50–57 (1999)

11. Iwata, T., Yamada, T., Sakurai, Y., Ueda, N.: Sequential modeling of topic dynamics with multiple timescales. ACM Trans. Knowl. Discov. Data **5**(4) (2012)

12. Karrer, B., Newman, M.E.J.: Stochastic blockmodels and community structure in networks. Phys. Rev. E **83**, 016107 (2011)

13. Kemp, C., Tenenbaum, J.B., Griffiths, T.L., Yamada, T., Ueda, N.: Learning systems of concepts with an infinite relational model. In: Proceedings of the 21st National Conference on Artificial Intelligence, Boston, Massachusetts, USA, vol. 1, pp. 381–388 (2006)

14. Klimt, B., Yang, Y.: Introducing the Enron corpus. In: First Conference on Email and Anti-Spam CEAS, Mountain View, California, USA (2004)

15. Kobayashi, T., Eguchi, K.: Online inference of mixed membership stochastic blockmodels for network data streams. IEICE Trans. Inf. Syst. **E97-D**(4), 752–761 (2014)

16. Miller, K.T., Jordan, M.I., Griffiths, T.L.: Nonparametric latent feature models for link prediction. In: Advances in Neural Information Processing Systems, vol. 22, pp. 1276–1284 (2009)

17. Nowicki, K., Snijders, T.A.B.: Estimation and prediction for stochastic blockstructures. J. Am. Stat. Assoc. **96**(455), 1077–1087 (2001)

18. Opsahl, T., Panzarasa, P.: Clustering in weighted networks. Soc. Netw. **31**(2), 155–163 (2009)

19. Snijders, T.A.B., Nowicki, K.: Estimation and prediction for stochastic blockmodels for graphs with latent block structure. J. Classif. **14**, 75–100 (1997)

Results of a Survey About the Perceived Task Similarities in Micro Task Crowdsourcing Systems

Steffen Schnitzer$^{(\boxtimes)}$, Svenja Neitzel, Sebastian Schmidt,
and Christoph Rensing$^{(\boxtimes)}$

Multimedia Communications Lab, Technische Universität Darmstadt,
Darmstadt, Germany
{Steffen.Schnitzer,Svenja.Neitzel,Sebastian.Schmidt,
Christoph.Rensing}@kom.tu-darmstadt.de

Abstract. Recommender mechanisms can support the assignment of jobs in crowdsourcing platforms. The use of recommendations can improve the quality and outcome for both worker and requester. Workers expect to get tasks similar to previously finished ones as recommendations, as a preceding study shows. Such similarities between tasks have to be identified and analyzed in order to create task recommendation systems that fulfil the workers' requirements. How workers characterize task similarity has been left open in the previous study. Therefore, this work provides an empirical study on how workers perceive the similarities between tasks. Different similarity aspects (e.g., the complexity, required action or the requester of the task) are evaluated towards their usefulness and the results are discussed. Worker characteristics, such as age, experience and country of origin are taken into account to determine how different worker groups judge similarity aspects of tasks.

Keywords: Crowdsourcing · Recommender systems · User survey

1 Introduction

Micro-task markets such as provided by the crowdsourcing platforms like *Amazon Mechanical Turk*[1] and *Microworkers*[2] publish tasks or campaigns of requesters in order to outsource projects over the Internet. Workers can search for tasks on the platform, submit a solution and receive a payment from the requester. In order to assign tasks, micro-task markets rely on the selection capabilities of the workers. The submitted solutions can be rejected by the requester, when the results are of unacceptable quality. A high rejection rate that is reported in this study and many others [7,10] shows that workers get assigned to task while being under-qualified. On the other hand we assume, that

---

[1] www.mturk.com.

[2] www.microworkers.com.

© Springer Nature Switzerland AG 2019
M. Atzmueller et al. (Eds.): MUSE 2015/MSM 2015/MSM 2016, LNAI 11406, pp. 107–125, 2019.
https://doi.org/10.1007/978-3-030-34407-8_6

also workers are assigned to tasks while being over-qualified. Therefore, both sides remain unsatisfied with parts of the results. Supporting the task selection process by helping workers to find matching tasks can be achieved by recommender systems. A task recommendation system has to match task requirements and the workers competences while also taking the expectations and preferences of workers into account. A preceding study has shown that tasks that are similar to recently finished ones are expected by the worker to be recommended by such a system [10].

We therefore consider similarities between tasks as a fundamental method in the design of task recommender systems for crowdsourcing platforms. However, how such task similarities have to be designed remains unclear. Therefore, this work provides qualitative and quantitative results from a survey on how workers perceive similarities between tasks.

The survey provides data on 500 submissions with equal contributions from 5 different regions of the world. Submissions that did not provide valid information were filtered with care and the survey design was specifically crafted in order to gain valuable insights from the data. We gathered results on the workers' general opinion on task recommendation and also included questions about the demographics of the workers. To determine the perceived task similarities, the participants had to judge 14 different similarity aspects. We provide the results about those similarity aspects in an overall evaluation, but also consider region dependent results, where significant differences between the regions were identified. Besides the evaluation of the 14 similarity aspects, the five most valued similarity aspects are used to analyze differing judgments depending on the workers' age, activity, payment and experience. Qualitative results provided by free text comments submitted within the survey are also provided.

The remainder of this paper is structured as follows. Section 2 introduces the current state of recommender systems for micro-task markets and task selection preferences in crowdsourcing platforms. Section 3 elaborates on the design and execution of the survey and the results are presented in Sect. 4. A summary, the conclusion of the paper and further research ideas are given in Sect. 5. This paper is an extended version of our paper [9] published at the MSM workshop in 2016. We added miscellaneous details and results of the study which have not been published in the MSM version.

2   State of the Art

Recommender systems follow either content-based or collaborative approaches. Content-based systems rely on the historic profile of the user to find similar tasks. Collaborative approaches rely on the preferences of similar users, while hybrid systems are also focus of research [3].

Following these methodologies, some approaches for task recommendation in crowdsourcing systems have been published in recent years. The "TaskRec" approach of Yuen et al. [12] was developed as a recommender system using collaborative approaches. They also include a user-category preference matrix to finally judge the worker's preference towards a certain task. Ambati et al. [1] describe a content-based recommender system, using the bag-of-word scheme to calculate similarities between tasks. Their system relies on a classification of all available tasks based on the history of a worker. Geiger [5] also describes a recommender system focused on the history of a single worker. He calculates a preference estimate by taking the requester and the keywords (tags or categories) of a task into account.

All of those approaches assume that similarity measures, which are based on bag-of-words, keywords or categories, suffice to model user behaviour. Therefore, we want to employ approaches that rely on the requirements gathered from the workers and explore possibilities of more sophisticated similarity measures, based on the task descriptions and taking the needs of the workers into account.

While focussing on the perceived similarities between tasks, our survey indirectly targets the motivation of the worker to choose a certain task. Insights to the motivation of workers in crowdsourcing systems have been presented already. Brabham et al. [2] discuss motivations to participate in crowdsourcing in a first place, while Kaufmann et al. [8] focus their study on the micro-task market *Amazon Mechanical Turk*. They distinguish the workers based on their demographic background and analyze the motivation of a worker from their task selection preferences. The search strategies and the task selection behaviour of workers have been examined by analyzing their task processing history or individual worker characteristics [4,6,11]. A preceding study shows that many workers expect, besides money and time related criteria, to get recommendations based on the similarity to the most recently completed task [10]. This survey provides a closer look on how to interpret this criterion of "similarity" between two tasks. Analyzing how workers perceive task similarities with a specific focus on recommender schemes is what distinguishes our work from the mentioned studies. This work provides insights on what kind of similarity measures are required in recommender systems for crowdsourcing platforms.

3   Methodology

3.1   Survey Design

The questionnaire is divided into four main parts shown in Fig. 1. The first part provides a motivation towards the ideas behind task recommendation for the participants' consideration. Here, the necessity of the survey is explained to the workers for them to understand our goals. To introduce the concept of recommender systems, this part gives examples on well-known product or content recommendations from online shopping or news platforms. This introduction also explains formalities like required time, assurance of confidentiality and exclusive scientific usage of the collected data. To make sure the workers take their time to read the introduction, we did not allow them to advance to the next part until at least one minute has passed.

The second part poses questions about the demographic background of the worker and the given experience within crowdsourcing platforms. The collected

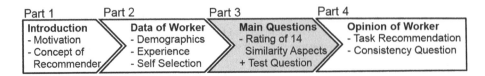

Fig. 1. The Design of the survey. How the different kind of questions are included in the four parts of the survey.

information is limited to gender, age, origin and residence of the workers. The experience with crowdsourcing was collected by including questions on how long and how intense such platforms have been used by the worker. The data about demographics and experience is only used for consistency checks of the workers answers and to check the quality of the submitted data. The analysis on worker characteristics was done using data taken directly from the crowdsourcing platform. Additionally, two questions asked about the worker's task search behavior to investigate how long workers search for tasks and whether they perceive that it is easy to find interesting tasks.

The third part of the survey holds the main part with questions focusing on the perceived task similarities. The 14 different similarity aspects are discussed below and given in Table 1. To detect spammers and filter invalid submissions a test question is introduced within this main part. The introduction of the questionnaire points out, that such kinds of attentiveness checks are included. Figure 2 depicts two example questions from the questionnaire, including the test question for attentiveness check.

After judging the similarity aspects of tasks, the workers are supposed to provide further opinions about task recommendation on the fourth part of the questionnaire. Some questions in part two and four are used as consistency questions [7], to be able to filter invalid submissions where the test question in the main part was answered correctly. In the main part of the questionnaire 14 similarity aspects have to be judged. The according questions all follow the same style. The test question, which is placed in the middle of this part, also follows this style. For each aspect a statement is given, where two tasks share the same value, e.g. "Task B takes the same time as Task A". The participants are advised to judge how useful this aspect is to determine the similarity of the two tasks. The answer is provided by selecting from a 5 point Likert scale between "not useful at all" and "very useful". Each of the questions also include a description with further advice or examples of the mentioned similarity aspect (see Fig. 2). The 14 aspects and their explanations are given below, for further details about the survey design please refer to the detailed description of the survey[3].

The different similarity aspects were chosen due to different properties and how they can be constructed from the given data. The aspects include six basic criteria of the tasks which are received as meta data. These *factual* attributes are

---

[3] http://www.kom.tu-darmstadt.de/~schnitze/files/msm_www16_survey.pdf.

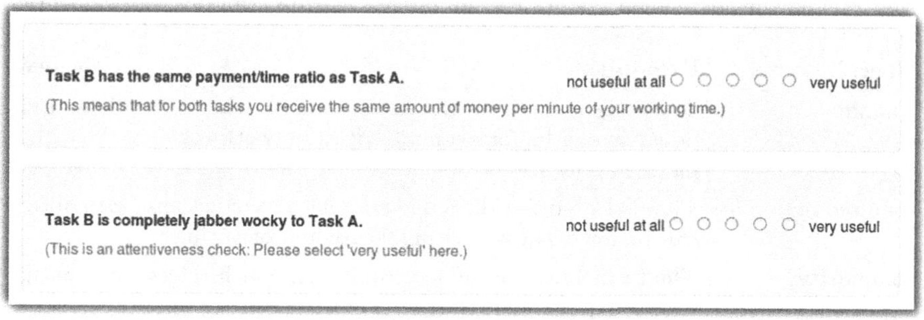

Fig. 2. Two example questions from the main part of the survey. The statement and description of the similarity aspect *payment* is shown at the top. The attentiveness check is shown at the bottom. On the right is the 5 point Likert scale.

related to payment and time of the task and have been shown to be very important for workers on crowdsourcing platforms [10]. Three more aspects consider the task's requester, which was also shown to be of interest for the workers [5]. Additionally, we introduce five similarity aspects which cannot be constructed from the meta data. These aspects are chosen to represent information which could be received from the task descriptions. Those *semantic* attributes are introduced to determine, whether such further similarity measures are required, which include information beyond the meta data of tasks.

On each page of the survey, an additional text field is given, where participants are able to provide general remarks. At the end of the main part, it is also possible to enter remarks about further possible aspects of similarity.

3.2 Survey Execution

The survey was executed between November 18th and December 4th 2015 and published on the micro-task market platform *Microworkers*. Another study suggests [10], that the region the workers are coming from, determine their task recommendation preferences. Five different regions were identified to be relevant. Many of the countries were chosen du to their high worker count on the platform (the top ten countries are included). We gathered 100 valid submissions per region. A defined "German speaking" region (AT, CH, DE) did not receive enough results to be considered and was replaced by the "Europe, West" region. An overview on the regions and the included countries is given in Table 2. The table also shows the amount of submissions per country of residence, as given in the questionnaire. The wage per region was decided upon a recommendation from the platform and previous experience with this crowdsourcing platform. Before 100 valid submissions were gathered, many invalid submissions had to be filtered out. Table 2 gives the spam rate, stating how many submissions had to be rejected until enough valid submissions were gathered. The "Europe, East"

Table 1. The considered aspects of task similarity, their descriptions and categories.

Aspect	Description	Category
Domain	The tasks come from the same domain, where domains can be for example social networks or mobile applications	Semantic
Required action	The tasks require the same actions, where actions can be for example writing, searching or voting	Semantic
Complexity	The tasks have the same complexity, which refers to the requirements needed for successful completion	Semantic
Comprehensibility	The tasks' descriptions have the same comprehensibility, which refers to the quality of language or the structure of the instructions	Semantic
Purpose	The tasks have the same purpose, where the purpose can be for example scientific or commercial	Semantic
Payment	The tasks have the same absolute payment, which means that a worker is being paid the same amount of money after successfully completion of the tasks	Factual
Time to finish	The tasks require the same time for completion	Factual
Payment/Time	The tasks have the same payment per time, which means that a worker receives the same amount of money per minute of his working time on the tasks	Factual
Time to rate	The tasks have the same time to rate, which is the maximal time the employer has to rate the task before it is rated satisfied automatically	Factual
Success rate	The tasks have the success rate, which is the ratio of submitted tasks rated as satisfied among all submitted tasks	Factual
Number of open tasks	The tasks' campaigns have the same number of open tasks left	Factual
Employer experience	The tasks' employers are registered for the same time or are equally active on the platform	Employer
Employer country	The tasks' employers have the same country of residence	Employer
Employer type	The tasks' employers have an equal type, where type can be for example commercial, scientific or well-known	Employer

region provided 100 valid submissions from 153 total submissions. This yields a spam rate of 35% for this region. We gathered 500 valid submissions in total, equally divided between the five regions to base our results on interpretable data.

Table 2. This table shows the number of valid submission per country for each region. Also the wage given for the survey and the detected spam rate for each region.

Region	Wage	Spam rate	Residence country (Code: ISO 3166-1)
Asia, South	$0.30	63%	BD(77), IN(13), LK(6), NP(3), PK(1)
Asia, South East	$0.30	52%	ID(34), MY(28), PH(27), VN(6), SG(3), TH(2)
English speaking	$0.50	35%	US(62), UK(21), CA(13), AU(4)
Europe, East	$0.40	35%	RS(34), RO(12), MK(12), BA(11), BG(10), HR(7), PL(4), LT(3), UA(2), TR(2), SI(1), HU(1), CZ(1)
Europe, West	$0.40	42%	IT(28), BE(19), FR(16), PT(14), ES(9), DE(6), FI(3), CH(2), IE(1), DK(1), AT(1)

The previously mentioned test question as attentiveness check and the consistency questions were used to filter the submissions accordingly. The attentiveness check was provided together in the middle of the main part of the questionnaire. This question, including non-existent words also stated: "This is an attentiveness check: Please select 'very useful' here". Still many of the workers failed to follow this instruction. The consistency questions gathered data about the experience of the worker. The answer to the question in part four can be derived from two questions in part two. Checking these answers for consistency as well as with data available from the platform allowed us to judge the validity of the submissions. In total, more than 47% of the submissions were invalid.

4 Results

This section provides the results of the survey. To show the attitude of the workers towards task recommendation, the results of the general questions are presented in Sect. 4.1. The overall results from the 500 gathered submissions are presented in Sect. 4.2. The differences between the regions is then analyzed in detail in Sect. 4.3. Demographic and other criteria are also used to analyze the results. Section 4.4 focuses on such criteria like the experience or the activity of the workers. The free text answers of the participants are considered in Sect. 4.6.

4.1 Is Task Recommendation Wanted?

In order to judge the workers' general acceptance of task recommendation and to understand what our participants expect from recommender systems, two

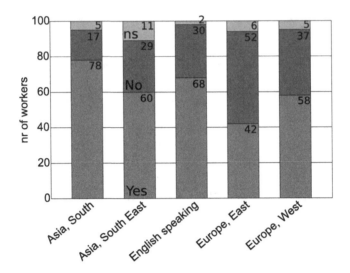

Fig. 3. The distribution of answers on whether it is easy to find enjoyable tasks per region. (ns = no statement)

questions were included in the survey. The workers were asked: "Do you think it is easy to find tasks that are interesting and enjoyable to work on?" and "Would you like to receive task recommendations on the platform?".

The first question, whether it is easy to find interesting enjoyable tasks, was positioned in the second part of the survey. Figure 3 shows the voting behaviour of the different regions. In "Asia, South" about 78% voted for "Yes" while from participants from the region "Europe, East" only 42% think it is easy to find enjoyable tasks. This difference of 36% points is the most significant between the regions. Overall, 61.2% think it is easy to find enjoyable tasks. 33.0% do not think that it is easy to find according tasks while 5.8% gave no statement (ns). This means that at least a third of the workers have difficulties to find appropriate tasks. Therefore we conclude, that further support for the task selection process is necessary. The second question was positioned in the fourth part of the survey. At this position, the participants were already exposed to the possible aspects of similarity between tasks. Here, 74.6% of the participants answered with "Yes" and the rest chose a negative answer.

The voting behavior, given in Fig. 4, shows differences of up to 25% points between the regions. Compared to the lowest results of 65% positive votes for task recommendation in the Englisch speaking countries, the participants from South East Asia had the highest results of 90% positive votes. Participants with a positive answer on this question had to answer a second question. They were asked: "Would you like to receive recommendations of tasks, which are similar to previously finished tasks?". By answering this question, the participants judged the similarity measure from the third part of the survey. Here the positive answers range from 97.2% to 98.9%. The throughout positive attitudes of

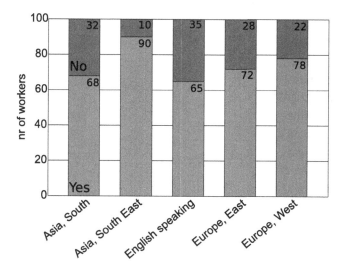

Fig. 4. The distribution of answers on whether recommendation is wanted in general (only Yes or No possible).

workers towards task recommendation based on similarities make a strong point. Such similarities need to be taken into account and analyzed in more detail for recommender systems in crowdsourcing platforms.

4.2 Overall Results

The main results on the judgement of the similarity aspects is given in Fig. 5. It shows the ratings for the different similarity aspects from the collective 500 submissions. The Likert scale answers are converted into ratings with values of '0' for "not useful at all" up to '4' for "very useful". For a detailed understanding of the results, we also give the mean of the weighted rating for each aspect (see Table 3). The aspects in Fig. 5 and in Table 3 are sorted from highest overall average rating to the lowest.

The overall result shows that the aspect of *action* is ranked the highest. This is the most important similarity feature for workers according to this survey. In general, all the aspects but one are rated with a positive tendency (with a mean above 2.0). Only the *employer country* is an exception to this. For each other aspect at least 70% of the participants judge the aspect as of having neutral to very positive value. More than 50% of the workers judge the first 11 of 14 aspects to be "useful" or "very useful". Observing similarly ranking aspects in Fig. 5, one can see that their voting behaviour appears to be also very similar. However, when the region of the worker is taken into account this picture of indifference is partly dissolved.

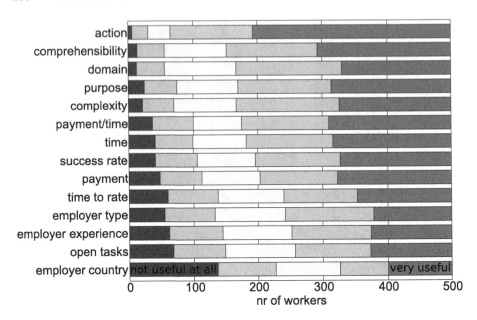

Fig. 5. The similarity aspects judged by 500 workers on a Likert scale from "not useful at all" to "very useful" (ordered by average rating).

In the following, the overall results of the regions is depicted, while Sect. 4.3 concentrates on the significant differences between the regions. The results within the regions show deviations from the overall results. The Asian regions differ the most from the common picture. Figure 6 shows the exact voting behaviour from "Asia, South". Participants from this region show a positive attitude towards the aspect of *action*, while aspects such as the *success rate* and employer related aspects are also valued high. Compared to the overall results, the tendency to get a positive answer is higher in this region. The most positive answers were gathered for the South East Asian region (see Fig. 7), where some of the aspects did not receive a single vote for "not useful at all". As the spam rate is very high in those two regions and our attentiveness check favours "very useful" answers, this appears suspicious. However, a more thorough data validation than described was not feasible. Also, the region with the highest detected spam rat does not provide the most positive answers. The most critical participants belong to the Western European region. Still, the *action* aspect did not receive a single "not useful at all" vote in this region. For the overall results the ratings were averaged over all corresponding submissions (see Table 3). The column *Overall Average* shows the overall average rating of each aspect and gives the rank in brackets. The similarity aspects are ordered from the highest to the lowest overall average rating. The detailed results for the different regions are given in the following columns in Table 3. The ranks of the similarity aspects are computed separately for each region and also given in the brackets. To easily find the five best ranked similarity aspects per region, they are given in bold font. None of

Table 3. The similarity aspects given with their average rating and their ranks in general and per region.

Similarity aspect	Overall average	Average Rating and (Rank) by Region				
		Asia S.	Asia S.E.	Engl. sp.	EU E.	EU W.
Action	**3.41 (1)**	**3.25 (1)**	**3.43 (1)**	**3.52 (1)**	**3.36 (1)**	**3.49 (1)**
Comprehensibility	**2.97 (2)**	2.96 (6)	**3.16 (2)**	**3.07 (2)**	2.79 (4)	**2.87 (2)**
Domain	**2.87 (3)**	2.92 (7)	**3.13 (3)**	2.73 (7)	**2.87 (2)**	**2.69 (4)**
Purpose	**2.83 (4)**	**2.99 (4)**	**3.05 (5)**	**2.84 (4)**	**2.74 (5)**	2.54 (6)
Complexity	**2.83 (5)**	2.74 (13)	2.97 (6)	**2.89 (3)**	**2.81 (3)**	**2.74 (3)**
Payment per time	2.76 (6)	2.83 (11)	2.83 (10)	**2.79 (5)**	2.73 (6)	**2.60 (5)**
Time	2.72 (7)	2.87 (8)	2.97 (6)	2.78 (6)	2.62 (8)	2.38 (7)
Success rate	2.66 (8)	**3.10 (2)**	**3.13 (3)**	2.40 (9)	2.47 (9)	2.20 (8)
Payment	2.63 (9)	2.84 (10)	2.88 (8)	2.53 (8)	2.71 (7)	2.17 (9)
Time to rate	2.42 (10)	2.81 (12)	2.83 (10)	2.26 (10)	2.08 (13)	2.11 (10)
Employer type	2.38 (11)	**3.02 (3)**	2.70 (13)	2.16 (11)	2.22 (10)	1.82 (11)
Employer experience	2.33 (12)	**2.97 (5)**	2.84 (9)	1.92 (13)	2.20 (12)	1.74 (12)
nr of open tasks	2.31 (13)	2.87 (8)	2.73 (12)	2.06 (12)	2.22 (10)	1.65 (13)
Employer country	1.82 (14)	2.60 (14)	2.37 (14)	1.38 (14)	1.41 (14)	1.34 (14)

the regions corresponds exactly with the overall results. Looking at the ranks of the similarity aspects, the European regions appear to be closest to the overall results, while the Asian regions appear to differ more. The similarity aspect *action* occupies the first rank in the overall analysis with an average rating of 3.41. In the region "Asia, South" it was voted to the first rank as well with a rating of 3.25. The similarity aspect *comprehensibility*, occupying the second rank in the overall rating, was voted to the sixth rank in South Asia, with a rating of 2.96. We described before, that the different similarity aspects are grouped into "semantic", "factual" and "employer related" aspects. The semantic similarities occupy the first five ranks of the overall rating. Most of the factual aspects follow afterwards and the employer related aspects can be found among the last ranks (with exception to the *Nr. of open tasks* aspect). This proves our hypothesis, that sophisticated similarity aspects, based on the task descriptions, are relevant for the workers on the platform.

On the other hand, each of the aspects accumulated many positive votes. Focusing on the aspect of *employer country*, which appears to be the least valued similarity aspect, this aspect still received 100 votes to be "very useful". This means 1/5th of the workers value this aspect very high, while less than half of the workers tend towards a negative response about this aspect.

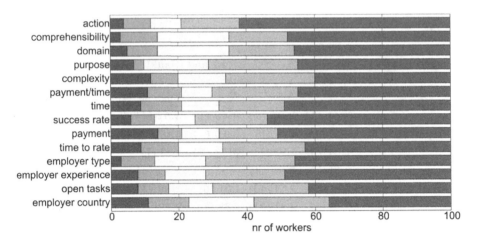

Fig. 6. The similarity aspects judged on a Likert scale from "not useful at all" to "very useful" by 100 workers from the South Asian region (ordered by overall average rating).

4.3 Results Depending on Region

Table 3 can be used to find the differences between the regions in detail. In the following, we focus on the most significant differences between the regions, which means that the data allows us to reject the null hypothesis that the samples come from the same population with at least $p < 0.05$. In the overall ranking, the aspects *success rate*, *employer type* and *employer experience* rank in places 8, 11 and 12 respectively. This is where the most differing region (South Asia) also shows the biggest differences in ranking those aspects to ranl 2, 3 and 5 respectively. Asian regions can be distinguished from the other regions, as their workers favour the *success rate* with a significance value of $p < 0.005$. The aspects of *open tasks*, *employer experience* and *employer type* follow this pattern of difference between Asian regions and the rest. Workers from the South East Asia region rate the similarity aspect *same domain* significantly higher than workers from English speaking countries or Western Europe. The similarity aspect *comprehensibility* is valued more by workers from the South East Asian region than by workers from Eastern Europe. Workers from the Asian regions rank the aspects *purpose*, *payment* and *time to rate* higher than participants from the region of Western Europe (even with $p < 0.001$). Workers from the Asian and the English speaking regions are significantly more interestes in the aspect of *time* than workers from Western Europe. For the aspects of *action*, *complexity* and *payment per time* no significant differences between the regions can be found.

4.4 Results Depending on Worker Characteristics

The results are also analyzed by considering worker characteristics like gender, age, activity, experience and payment. The information about these characteristics were gathered from the platform and have not been part of the questionnaire.

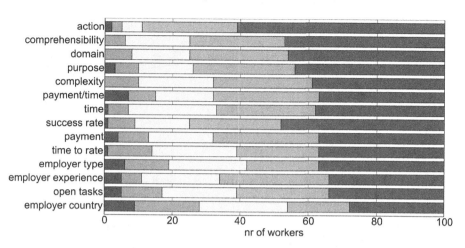

Fig. 7. The similarity aspects judged on a Likert scale from "not useful at all" to "very useful" by 100 workers from the South East Asian region (ordered by overall average rating).

Table 4. The quartile distributions for the different worker characteristics.

Quartile	Age (years)	Activity (task/day)	Payment (USD)	Experience (tasks)
1	<23	<0.450	<$0.110	<57
2	<27	<1.611	<$0.138	<387
3	<35	<4.070	<$0.200	<1809
4	<69	<46.040	<$5.274	<50322

It was tested if there are significant differences in the rating of similarity aspects for different genders. Of the 500 participants of all five regions, 328 are male and 171 female. The questionnaire offered a third option *other*, which is not considered here due to its limited occurrence. Although the ranking of similarity aspects is slightly different for male and female workers, there were no significant differences found. The most differently rated similarity aspect is *complexity* with an average rating of 2.76 for male and 2.96 for female workers. However, this difference does not reach statistical significance ($p < 0.06$).

The *age* of a worker is given by date of birth. The *activity* of a worker can be described as "more tasks done in less time" and is calculated by dividing the number of tasks by the membership time. The *experience* of a worker is given as the overall number of tasks the worker submitted. The average *payment* of a worker is calculated by dividing the overall earnings by the overall number of tasks done. For each characteristic the population was split by quartiles into four equally sized sub samples described in Table 4. This was done for the whole set of submissions, leaving 125 submissions in each part.

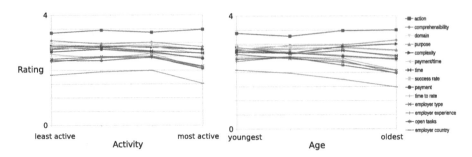

Fig. 8. Average aspect ratings for overall quartiles.

Figure 8 shows the overall average ratings of similarity aspects for different quartiles of the characteristics activity and age. For the activity characteristic, the visualization suggests that more active workers rate aspects other than *action* lower than less active workers. For the age characteristic, there is a broadening of ratings visible for increasing age. That means that the range of means of the similarity aspects changed from (2.09–3.38) in the first quartile to (1.48–3.5) in the last quartile, which shows that the means are more spread throughout the range in the last quartile. This could indicate that older workers give more distinct ratings for the different aspects. When it comes to activity, the more active workers tend to generally rate most of the similarity aspects lower than the less active workers. For most of the other criteria, there was no clear image on how those characteristics influence the perception of similarity aspects. A further analysis, which determined the characteristics dependent on the region was done and the results are given in the next section.

4.5   Results Depending on Worker Characteristics and Region

Analyzing the results for influences of certain worker criteria does not yield many interesting insights. However, as the different regions were found to be very diverse in their answers, an analysis that determines the worker characteristics separately for each region leads to more detailed conclusions. For the reason of simplicity and the focus of our research, the results in this section present the five most valued similarity aspects only.

The submissions for each region were again split into four quartiles, leaving 25 submissions for each quartile per region. The quartiles were generated separately for each region and each worker characteristic. This results in differently defined worker quartiles, where e.g. the first quartile for the characteristic *age* contains submission from workers between 17–23 years in the English speaking region, and submission from workers between 16–22 in the South East Asian region.

Comparing the most diverse results for the worker characteristic age, the results from Western Europe are compared to the results from South Asia. With increasing age, the Western Europe region shows an increase for all of the chosen aspects, while the South Asian region mostly shows a decrease in how high the

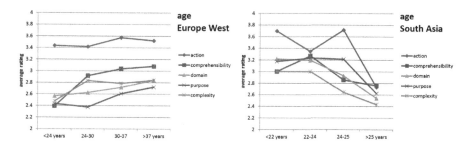

Fig. 9. Average ratings per aspect for the worker characteristic *age* in the regions of Eastern Europe and South Asia.

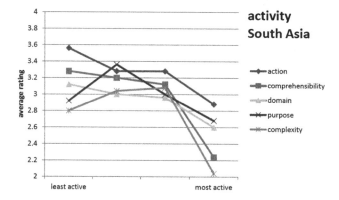

Fig. 10. Average ratings per aspect for the worker characteristic *activity* in the South Asian region. From least active workers on the right to most active workers on the left.

aspects are valued. This might be due to a different perception of the aspects within the two different regions. However, the quartiles show, that in South Asia many more people in the age range up to 25 are represented (about 2/3) while workers from this range contribute less than half of the submissions for the Western European regions. In general, the Western European region is more representative for the other regions, according to the age range and the increased perceived value of these aspects in "older" quartiles. The worker characteristic *activity* does not reveal certain patterns for most of the regions. The most interesting results can be found for the South Asian region. The results are depicted in Fig. 10 and show a significant decrease in the average rating of all the similarity aspects. This effect was also observed for the other similarity aspects not shown in the figure. More active users in this region tend to value all of the aspects less than workers that are less active. This can be due to the fact, that more active users know their ways on the platform very well, and therefore judge the necessity of a recommender system to be less in general. It is also possible, that more active users in the South Asian region are more critical towards the different aspects (Fig. 10).

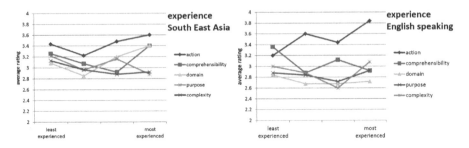

Fig. 11. Average aspect ratings for the worker characteristic *experience* in the South East Asian region and the English speaking region from least experienced workers on the left to the most experienced workers on the right.

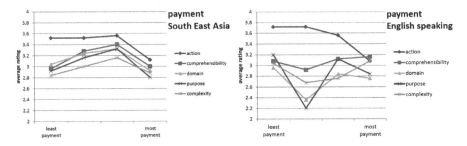

Fig. 12. Average aspect ratings for the worker characteristic *payment* in the South East Asian region and the English speaking region from workers with the least payment on the left to workers with the most payment on the right.

The worker characteristic *experience* categorizes the workers depending on the number of tasks they submitted on the platform. Focusing on the similarity aspect *action* there are different developments to be found for the regions. In Eastern Europe as well as in South Asia, the average rating for *action* decreases towards more experienced workers. However, in Western Europe as well as in the two depicted regions (see Fig. 11) more experienced users value this similarity aspect higher. The different behaviour for the other aspects shows once more, that the perception of the similarity aspects differs substantially between the analyzed regions. But also within the regions, the data does not allow profound conclusions drawn from the experience of the workers. This might show, that the perception of the aspects is not necessarily dependent from the experience of workers. On the other hand, the definitions of the regions might be too broad and a more detailed analysis per country is necessary to show how these characteristics influence the perception of similarity aspects.

The last worker characteristic that was analyzed is the *payment*. Here the average wage for each worker was calculated and used to sort them into the quartiles. Again the behaviour is very different between the regions. However, for some regions patterns can be found in the data. In the South East Asian region (see Fig. 12) all the shown aspects increase towards the third quartile but suddenly decrease in the last quartile. This behaviour shows some parallels to the activity characteristic in the South Asian region shown in Fig. 10. It is possible, that this reveals a certain user group within this region, that is paid well and that shares a common perception towards the usability of the similarity aspect. The results for the English speaking region (see Fig. 12) on the other hand show no common pattern for the similarity aspects. However, the aspect *action* shows a significant decrease towards higher paid workers. The perception between the four quartiles differs very much while most of them share a decrease from the first to the second and an increase from the second to the third quartile. This can also be an indicator, that the worker characteristic of *payment* is a good feature to divide different worker groups that share common perception in contrast to the other groups.

Overall, the analysis of the characteristics within the different regions showed, that patterns can be found within the data, and that the given characteristics allow certain conclusions. For a task recommender system, this means that the origin of a worker is a valuable feature to base recommendations on. Also, in some cases, the presented worker characteristics allow to conclude towards certain preferences of workers in regards of task similarity. In general, the given similarity aspects are a good starting point, however, the results also show that the perception and preferences may differ on a more fine granular level than what was able to show with the provided data.

4.6 Insights from Free Text Comments

In the questionnaire workers were given the opportunity to provide own ideas for further aspects of task similarity. They suggested the form of the proof to submit after finishing a task as a further aspect. This could be useful to workers because a transparent verification procedure (e.g., a verification code) ensures that their work will be recognized properly and they receive their payment quickly. Another aspect suggested by the workers was the destination a task forwards the worker to in order to complete it. Maybe there are some platforms the user is not able or willing to access (e.g., social networks where an account is needed) or the worker is already familiar with some platforms and therefore able to finish tasks on these platforms faster. Some workers also mentioned the requirement of specific skills as another aspect to determine task similarity.

Workers were further given the opportunity to mention other recommendation criteria, apart from task similarity, that seem important to them. Many criteria were given like best payment, shortest time to finish, shortest time to rate, best payment per time and best success rate. Many workers suggested recommending tasks based on the skills of a worker. Another frequent comment was the availability and proper functioning of destination platforms as many tasks seem to not work correctly. Some workers want to receive recommendations of tasks from employers they already worked for.

Some suggestions for general improvements of Microworkers were also given: One worker proposed to introduce a color pattern for indicating the difficulty of tasks and another suggested that more complex and better paid jobs should be offered to workers with a higher rate of satisfying tasks. Another worker proposed the possibility to mark tasks as unwanted and to rank similar tasks lower in the list of available jobs. Others suggested to introduce a possibility for workers to rate employers and single tasks.

5   Conclusion and Outlook

In this work we analyze requirements of workers on a crowdsourcing platform towards task recommendation. We motivate and describe the design and execution of a survey to achieve this goal. Our results allow the conclusion, that task recommendation is welcomed by the workers. The results clearly indicate, that the perceived similarity between two tasks depends on many different aspects. However, the opinions of the workers differ depending on the different regions of residence of the workers.

We want to emphasize that the aspects that we defined as *semantic* occupy the first ranks in the overall analysis. According to the precedent study [10] these similarity based criteria are valued less than money related criteria. Analyzing the perceived similarity of tasks in this work leaves money and time related similarity aspects to be rated lower. This shows that such factual task characteristics are highly relevant for recommendation, but the characteristics ranked highest in this survey are more relevant for recommendations based on the similarity of tasks.

Considering the differences between the regions, it is very interesting, that the single aspect of *action* was rated to be the most important aspect throughout all the regions. Also, the most significant differences were found between the Asian regions and other regions.

The evaluation of the worker characteristics in detail, depending on the workers region gave some interesting insights as well. The characteristic on how much the workers are paid in average may be an indicator on how they perceive similarity measures and how their preferences for task recommendation systems are. The differences and also the commonalities found between the regions and also between the different worker quartiles shows how this data can be used for personalized task recommendation.

Overall, the results of this survey allows the conclusion, that sophisticated similarity measures are required for task recommendation in crowdsourcing systems. Workers from different regions agree to a certain extent, that the semantic similarity aspects (e.g. the *required action*) are more important to judge the similarity between tasks, than factual aspects like time and money. Although, the perceived task similarities varies significantly between different world regions, which shows, that task recommendation has to be personalized and go beyond the approaches proposed so far. In future work it will be necessary to examine the possibilities of using semantic similarity features derived from task descriptions in order to build task recommendation schemes. Also, classifying or clustering tasks depending on such similarities may help to improve existing task recommendation approaches which currently use categories provided by the platform or employers.

Acknowledgements. This work is supported by the Deutsche Forschungsgemeinschaft (DFG) under Grants STE 866/9-2, RE 2593/3-2, in the project "Design und Bewertung neuer Mechanismen für Crowdsourcing".

References

1. Ambati, V., Vogel, S., Carbonell, J.: Towards task recommendation in micro-task markets. In: Proceedings of the 11th AAAI Conference on Human Computation, AAAIWS 2011, pp. 80–83. AAAI Press (2011)
2. Brabham, D.C.: Moving the crowd at threadless: motivations for participation in a crowdsourcing application. Inf. Commun. Soc. **13**(8), 1122–1145 (2010)
3. Chartron, G., Kembellec, G.: General introduction to recommender systems. In: Kembellec, G., Chartron, G., Saleh, I. (eds.) Recommender Systems, pp. 1–23. Wiley, Hoboken (2014)
4. Chilton, L.B., Horton, J.J., Miller, R.C., Azenkot, S.: Task search in a human computation market. In: Proceedings of the ACM SIGKDD Workshop on Human Computation, pp. 1–9. ACM (2010)
5. Geiger, D.: Personalized Task Recommendation in Crowdsourcing Systems. Progress in IS. Springer, Cham (2016). https://doi.org/10.1007/978-3-319-22291-2
6. Goodman, J.K., Cryder, C.E., Cheema, A.: Data collection in a flat world: the strengths and weaknesses of mechanical turk samples. J. Behav. Decis. Mak. **26**(3), 213–224 (2013)
7. Hoßfeld, T., et al.: Best practices for QoE crowdtesting: QoE assessment with crowdsourcing. IEEE Trans. Multimedia **16**(2), 541–558 (2014)
8. Kaufmann, N., Schulze, T., Veit, D.: More than fun and money. Worker motivation in crowdsourcing-a study on mechanical Turk. In: AMCIS, vol. 11, pp. 1–11 (2011)
9. Schnitzer, S., Neitzel, S., Schmidt, S., Rensing, C.: Perceived task similarities for task recommendation in crowdsourcing systems. In: Proceedings of the 25th International Conference Companion on World Wide Web. International World Wide Web Conferences (2016)
10. Schnitzer, S., Rensing, C., Schmidt, S., Borchert, K., Hirth, M., Tran-Gia, P.: Demands on task recommendation in crowdsourcing platforms - the worker's perspective. In: ACM RecSys 2015 CrowdRec Workshop, Vienna (2015)
11. Yuen, M.-C., King, I., Leung, K.-S.: Task recommendation in crowdsourcing systems. In: Proceedings of the First International Workshop on Crowdsourcing and Data Mining, pp. 22–26. ACM (2012)
12. Yuen, M.-C., King, I., Leung, K.-S.: TaskRec: a task recommendation framework in crowdsourcing systems. Neural Process. Lett. **41**, 1–16 (2014)

Provenance of Explicit and Implicit Interactions on Social Media with W3C PROV-DM

Io Taxidou[1][(✉)], Tom De Nies[2], and Peter M. Fischer[1]

[1] University of Freiburg, Freiburg, Germany
{taxidou,peter.fischer}@informatik.uni-freiburg.de
[2] Ghent University - imec - IDLab, Ghent, Belgium
tom.denies@ugent.de

Abstract. In recent years, research in *information diffusion* in social media has attracted a lot of attention, since the data produced is fast, massive and viral. The *provenance* of such data is equally important because it helps to judge the relevance and trustworthiness of the information enclosed in the data. However, social media currently provide insufficient mechanisms for provenance, while models of information diffusion use their own concepts and notations, targeted to specific use cases. In this paper, we present and extend our model for information diffusion and provenance, based on the *W3C PROV Data Model*. PROV provides a Web-native and interoperable format that allows easy publication of provenance data, and minimizes the integration effort among different systems making use of PROV. We provide computational methods for provenance reconstruction of user interactions based on the investigation of human behaviour on social media.

Keywords: Provenance · Information diffusion · User interactions · PROV-DM

1 Introduction

Social media such as online social networks (e.g., Facebook), micro-messaging services (e.g., Twitter) or sharing sites (e.g., Instagram) provide the virtual space in which a significant part of social interactions takes place. Many real-life situations, such as elections, are reflected by social media. In turn, social media shape these situations by forming opinions or strengthening trends, or by spreading reports on emerging situations faster than conventional media. Furthermore, word of mouth plays an important role in shaping user's attitudes and behavior. Most importantly, social media provide a huge audience (some users maintain millions of connections) where information can be easily spread and consumed by others. This phenomenon is referred as *information diffusion.*

Due to the plurality of opinions and multiple sources of information in social media, the need for judging the relevance and trustworthiness of such information

© Springer Nature Switzerland AG 2019
M. Atzmueller et al. (Eds.): MUSE 2015/MSM 2015/MSM 2016, LNAI 11406, pp. 126–150, 2019.
https://doi.org/10.1007/978-3-030-34407-8_7

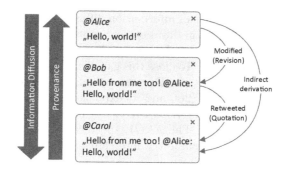

Fig. 1. Information diffusion and provenance

is becoming urgent. Understanding how a piece of information propagated in social media provides additional context, including the source and its properties, the intermediate forwarders and the modifications that this piece of information has undergone. A social media user (for example an online journalist) can take advantage of this context to assess the source and intermediate forwarders, to determine the impact of their own publications and also to predict information *virality*. Additionally, the detection of rumors is feasible not only by discovering the sources but also by analyzing the properties of the diffusion process [21] and the intermediate steps. When it comes to massive amounts of negative opinions expressed in social media, companies, politicians and celebrities need to react promptly by understanding who is propagating certain information and who is influencing others.

Such kind of analyses refer to the inverse process of information diffusion, *information provenance* that seeks the paths back to the sources. While provenance is a well researched topic in domains like workflows [14] or databases [6], it has received limited attention in the context of social media. Likewise, existing models of information diffusion are insufficient to model provenance, while the current structure of social media provides limited or no mechanism to its users to judge received information [4]. For example, in cases of retweets on Twitter, only the source of information is provided but not the intermediate steps (forwarders). However, it has been shown that forwarders play an equally important role in the outcome of information diffusion [3].

To further clarify the relation between information diffusion and provenance, we provide an example in Fig. 1. Three Twitter users are emitting a similar message: Alice is the source of information diffusion, as she emits an original message. At a later point in time, user Bob modifies the original message and then user Carol copies and forwards (retweets) the message of Bob. In this process, it is important to understand how the message was modified and forwarded. User Carol was indirectly influenced by user Alice, since her message was *indirectly derived* from the source (two-step procedure). This means that the trustworthiness of all three users involved should be judged, since they participate in the diffusion and modification of this message.

Despite the variety of models of information diffusion, there exists no unified, conceptual model for information diffusion and provenance that can be applied to different datasets and set-ups, while remaining both expressive and generic enough to cover many use cases. To close this gap, we presented the *PROV-SAID* model in [29], a model to assert the **Prov**enance of **S**ocial medi**A I**nformation **D**iffusion based on *PROV-DM* [26], and its extensions. *PROV-DM* is the main component of the wider family of *W3C PROV* documents that defines an interoperable model for provenance. The *PROV* specification provides the concepts and supporting definitions to enable the interchange of provenance information in heterogeneous environments such as the Web.

PROV-DM has many benefits: concepts of information diffusion and provenance are modeled in a *W3C* standard that allows subsequent integration and interaction with other tools that make use of PROV. As a result, the cost of integration is lowered since data derived from this modeling is exposed in a Web-native and interoperable format. This is very useful in cases where data needs to be combined from different (social media) sources that do not share the same concepts and notations. Additionally, *PROV-DM* is domain-agnostic, but it has the benefit of extensibility, allowing domain-specific information to be included.

The actual computation of provenance bears many challenges for the existing social media platforms. Information diffusion and provenance mechanisms are mainly studied using one obvious propagation means – e.g., retweets in Twitter, reposts in Facebook, revines in Vine. The combination of different means of propagation has not been investigated thoroughly, resulting in incomplete provenance. Implicit diffusion mechanisms adopted by social media users throughout well-known conventions are also understudied. The same is observed where users are influenced by others but no provenance indication is provided by them or by social media providers: often such valuable information is lost. We have empirically studied different *explicit* and *implict* propagation means and their correlations presented in [32]. However, such observations have not been fully evaluated. Taken together here, we evaluate and extend our methods for computing provenance when explicit [30], implicit [32] or no information [11] is provided.

This paper contains the following contributions: (i) We extend our previous work [29] by modeling further types of interactions in the *PROV-SAID* model: this way we consider all types of user interactions, including those we have empirically observed, not restricted to typical information diffusion. (ii) We present and unify our previous methods for computing provenance [11,32] and provide an evaluation that demonstrates multiple types of user interactions and influence in social media. We also present methods to represent the newly defined interactions in the *PROV-SAID* model.

The rest of the paper is structured as follows: in Sect. 2, we present the extended *PROV-SAID* model; in Sect. 4 we elaborate into the concept of Influence and its extensions; in Sect. 5 we present and unify our previous methods for provenance computation and in addition, we present methods for the newly defined *PROV-SAID* extensions; Sect. 6 presents an evaluation in Twitter and shows the applicability of our modeling and computational methods; in Sect. 7 we describe related work and Sect. 8 concludes the paper.

2 Model Overview

In Sects. 2, 3 and 4, we describe our model with its relevant extensions and constraints[1]. *PROV* has formal semantics [7], which cover our model as well, since our extensions and constraints are fully compliant with *PROV*.

Throughout the text, we provide a full example that covers all aspects of the model. To improve clarity, this example is unfolded incrementally and the reader should take into account information in previous examples.

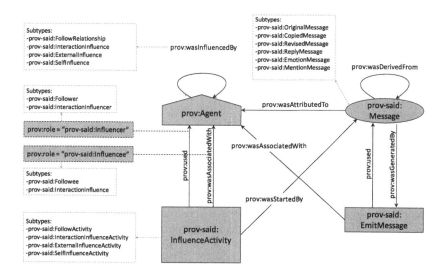

Fig. 2. *PROV-SAID* model

2.1 Overview

The *PROV-SAID* model can be applied to any social network where information propagates from user to user in the form of messages. Messages can be transmitted though *social connections*, but the model is general enough to capture *external influence* as well, as often happens in social media [27]. For example, Twitter users might publish information that has been seen on the public timeline without any direct social connection. Furthermore, our experience with provenance in Twitter shows that information does not flow only from social connections, but there is approximately 20% of external influence [31]. The last observation derives from experiments with reweets where diffusion is explicit,

---

[1] For detailed specification and formal constraints, see http://semweb.datasciencelab. be/ns/prov-said/.

while this percentage is much higher for non explicit diffusion (propagation of Twitter hashtags).

Our model includes *activities* and *relationships* connected with information diffusion, such as exchanging *messages*, finding the *source* of diffusion, and expressing which *changes* the message has undergone through this procedure. User *influence* plays a key role in information diffusion since it drives information flow. The concept of influence is only vaguely defined in *PROV-DM*. It is recommended to use more specific terms when possible, since influence can take many forms in different use cases. However, for our use case the influence relationship has its own merit. Therefore, we define and extend the concept of influence, expressed through different activities, types and user roles.

Figure 2 shows a high-level overview of the *PROV-SAID* model with its extensions. Throughout this work, the prefix *prov:* refers to the PROV namespace[2] and the prefix *prov-said:* refers to the new *PROV-SAID* namespace[3]. Users who emit messages on social media are represented by the *prov:Agent* concept. Terms *agent* and *user* are utilized in an alternate way: agent is mostly prov-said model specific, while user is a general term to denote participation in social media.

2.2   Design Decisions

The purpose of *PROV-SAID* is to offer an easily reusable model that covers and infers different aspects of information diffusion and provenance. The goal is not minimizing the relationships in the model, but offering maximum expressiveness.

Also, we aim at borrowing the already defined concepts from *PROV-DM* wherever possible, and defining our own extensions for specific use cases. This way we improve clarity and we encourage reusability of the model. One example of extending the model is the concept of *prov:Influence*: We differentiate the cases in the context of information diffusion and provenance and we provide a more clear meaning to them.

The *PROV-SAID* model refers to information diffusion and provenance in the context of social media; since social connections are the main carriers of information [31] we need to specify whether a message was propagated through them or whether there was some external influence. For this purpose, we implicitly model social graph connections (unidirectional relationships); additionally, the activity of connecting to others constitutes an interaction and shows some kind of influence. We proceed by describing the components and relationships of our model step by step.

3   Modeling Messages

In order to model messages that are emitted by users, we propose the following extensions that are subtypes of *prov:Entity*:

---

[2] http://www.w3.org/ns/prov.
[3] http://semweb.datasciencelab.be/ns/prov-said/.

– *prov-said:Message*: denotes the general class of messages.

Messages in social media might be original messages, copied messages or revised messages. We define the following categories as subtypes of *prov-said:Message*:

- *prov-said:OriginalMessage* denotes an original message that is not derived from any other message and the user who emitted it is the initiator of information propagation for a specific message.

- *prov-said:CopiedMessage* denotes a message which is based on another message that has been published in the past and was forwarded as an exact copy (such as the *retweet* function of Twitter). Users who emit copied messages comply fully with the content and opinions of the original message.

- *prov-said:RevisedMessage* denotes a message that is produced by modifying an existing message. This means that the user who emits such a message may or may not share the original opinion of the original message. It is possible that the information carried by the original message is altered.

- *prov-said:ReplyMessage* denotes a message that is produced by replying to an existing message. A reply is expressed normally by social media providers through an embedded function and often the name of the user who posted the existing message is mentioned. Note here that unless users explicitly annotate replies, it is hard to trace their provenance since there might be no other characteristics indicating their connection to the existing message.

- *prov-said:EmotionMessage* denotes an empty message that expresses emotions to an existing message. We decided to model it as an empty message because in most platforms the existing message appears in the timeline of the user who expressed an emotion towards it. For example, in Twitter a message is *favorited*, in Facebook *liked*, while recently users can express more emotions like anger, surprise, sadness, etc.

- *prov-said:MentionMessage* denotes a message that mentions another user. Such a message might be complementary to any type of the above mentioned. For example, a *revised message* might mentions another user. We have empirically identified the following semantics of a mention: (1) a user mentions another popular user (e.g. celebrity, politician); (2) a user mentions another user to expose information to them; (3) a user tags another user in a message to express a shared participation (e.g. being physically at the same place). We do not model such cases separately due to the difficulty of their detection.

Note here that the types *prov-said:OriginalMessage*, *prov-said:Copied-Message* and *prov-said:RevisedMessage* are strictly disjoint. In contrast, *prov-said:ReplyMessage* and *prov-said:EmotionMessage* are only disjoint with the type *prov-said:OriginalMessage*. For example, we have observed on Twitter that a *prov-said:ReplyMessage* can also be a *prov-said:CopiedMessage*. Finally, *prov-said:MentionMessage* can be complementary to any other type of message type, including *prov-said:OriginalMessage*. The focus of the latter type of *prov:Message* lies in the interaction of *prov:Agents*, rather than *prov:Messages*.

With these six cases, we have covered the main cases of information diffusion and interactions through messages that we see in the empirical data.

3.1   Message Attribution

A *prov-said:Message* is always attributed to a user *prov:Agent* using the relationship *prov:wasAttributedTo*. Example 1 illustrates the use of messages and attribution for the Twitter social network.

Example 1: Message Creation and Attribution

```
prefix twitter: <http://twitter.com/>
prefix alice-status: <http://twitter.com/Alice/status/>
prefix bob-status: <http://twitter.com/Bob/status/>
prefix carol-status: <http://twitter.com/Carol/status/>

// User @Alice tweeted a message "Hello, world!"
prov:entity(alice-status:12345, [prov:type='prov-said:OriginalMessage',
    prov:label='Hello, world!'])

// User @Bob modified and re-emitted the "Hello, world!" message
prov:entity(bob-status:23456, [prov:type='prov-said:RevisedMessage',
    prov:label='Hello from me too! MT @Alice: Hello, world!'])

// User @Carol retweeted (copied) the revised message
prov:entity(carol-status:34567, [prov:type='prov-said:CopiedMessage',
    prov:label='Hello from me too! MT @Alice: Hello, world!'])

// User @Dan replied to Alice's message
prov:entity(dan-status:45678, [prov:type='prov-said:ReplyMessage',
    prov:label='Hi @Alice!'])

// User @Eve favorited Alice's message
prov:entity(eve-status:56789, [prov:type='prov-said:EmotionMessage',
    prov:label='Favorite'])

// User @Frank mentioned Alice in his message
prov:entity(frank-status:67891, [prov:type='prov-said:MentionMessage',
    prov:label='@Alice just said hello'])

// alice-status:12345 was emitted by twitter:Alice
prov:wasAttributedTo(alice-status:12345, twitter:Alice)
```

3.2   Message Emission

Next we define the following activity that refers to message emission and is a subtype of *prov:Activity*

- *prov-said:EmitMessage* denotes a generic emission of a message. It must generate a *prov-said:Message*, and may use another *prov-said:Message*.

Note that the subtype of the generated *prov-said:Message* can be inferred from the usage of another *prov-said:Message* by the *prov-said:EmitMessage*. For example, if the content of the generated message is identical to that of the used one, it is a *prov-said:CopiedMessage*. The same applies for the other types of *prov:Message*s.

3.3   Message Derivation

Whereas an original message does not have dependencies on other messages, copied, revised, reply and emotion messages can be traced back to their original sources through derivation - i.e., they cannot exist on their own. *PROV-DM* already provides the basics for the concepts needed to model copied and revised messages, in the form of *prov:Quotation, prov:Revision,* and *prov:PrimarySource*, as illustrated by Example 2. To cover the newly defined interactions, we add the types *prov-said:Reply* and *prov-said:Emotion* as subtypes of *prov:Derivation*.

Example 2: Message Derivation

```
// bob-status:23456 was derived from alice-status:12345,
// which is also its primary source (in the context of Twitter)
prov:wasDerivedFrom(bob-status:23456, alice-status:12345, emit-23456, gen-23456,
   use-12345, [prov:type='prov:Revision', prov:type='prov:PrimarySource'])

// carol-status:34567 was quoted from bob-status:23456
// (which is not its primary source)
prov:wasDerivedFrom(carol-status:34567, bob-status:23456, emit-34567, gen-34567,
   use-23456, [prov:type='prov:Quotation'])

// dan-status:45678 was replied to alice-status:12345
prov:wasDerivedFrom(dan-status:45678, alice-status:12345, emit-45678, gen-45678,
   use-12345, [prov:type='prov:Reply'])

// eve-status:56789 expressed an emotion towards alice-status:12345
prov:wasDerivedFrom(eve-status:56789, alice-status:12345, emit-56789, gen-56789,
   use-56789, [prov:type='prov:Emotion'])
```

We observe that *carol-status:34567* was indirectly derived from *alice-status:12345*. To model this dependency, we introduce the concept *prov-said:IndirectDerivation* as a subtype of *prov:Derivation*. This way we can model multi-step provenance and trace how messages are being derived, without being restricted to the previous step only. We illustrate this in Example 3.

Example 3: Indirect Derivation

```
// carol-status:34567 was indirectly derived from alice-status:12345
prov:wasDerivedFrom(carol-status:34567, alice-status:12345,
   [prov:type='prov-said:IndirectDerivation'])
```

Note that the *prov-said:MentionMessage* might not be derived from another *prov:Message*. As a result, we do not need to express any additional sub-type of *prov:Derivation*. In case a *prov-said:MentionMessage* is derived from another one, this will be covered by the other defined sub-types of *prov:Derivation* (e.g. *prov:Quotation*).

At this point, we express the following constraints:

- An *prov-said:OriginalMessage* cannot be derived from a *prov-said:Message*.
- A copied, revised, reply, emotion message should always be derived from another message. A *prov-said:EmitMessage* activity that generates a *prov-said:CopiedMessage* and uses a *prov-said:Message* implies that the first message was derived from the latter by means of *prov:Quotation*. Analogously, generation of a *prov-said:RevisedMessage* and usage of a *prov-said:Message* by a *prov-said:EmitMessage* implies that the first message was derived from the later by *prov:Revision*. The same applies for *prov-said:ReplyMessage* and *prov-said:EmotionMessage* with the derivation types *prov-said:Reply* and *prov-said:Emotion*.

In the next Section we continue with modeling Influence with regard to information diffusion.

4   Modeling Influence

When users (*prov:agents*) are interacting by exchanging and forwarding messages, expressing emotions or mentioning other users, influence is created among them. Recall the definition of [26], where influence denotes to have an *effect* on something. Once message *prov:Derivations* are expressed (we will discuss in Sect. 5 how to compute them), we can also express who influences whom by identifying the authors of the messages. Note the exception of *prov-said:MentionMessage*, where influence is created instantly among the two agents involved (The existence of an influencing message is not necessary).

Influence plays a key role since it drives information flow in social media; however, it is challenging to capture its semantics and compute its sources. Users often react to messages by giving credit to the original contributor. However, they are also frequently influenced by external factors such as traditional media, and react in social media [27]. Influence has many ways of expression: through establishing social connections, interacting or expressing emotions over messages, mentioning other users, or reacting as a result of real events. In this Section, we propose extensions for influence types, influence activities and influence roles on top their generic and rather vague definitions in *PROV-DM*. In Sect. 5 we propose methods to compute in practice such influence.

4.1   Influence Types

We define a relationship: *prov-said:InfluenceRelationship* to express general influence and four subtypes to specify the ways that such influence can be expressed. Note here that by the term *influence* we do not refer to the classical influence expressed in state of the art for social media analysis, but we rather follow the general definition of *PROV-DM* [26] *"Influence is the capacity of an entity, activity, or agent to have an effect on the character, development, or behavior of another by means of usage, start, end, generation, invalidation, communication, derivation, attribution, association, or delegation."* This way all social interactions are captured: for example reply messages that do not demonstrate the typical social media influence, but an existing message is *having an effect* on the reply message which shows the *PROV-DM* influence.

Note that in addition, our model allows the representation of social connections among users, since information flows through them in the majority of cases. Furthermore, the model is generic enough to assert the provenance of information diffusion even without the presence of social connections.

First, we define a generic concept *prov-said:InfluenceRelationship* as a subtype of *prov:Influence*. It denotes an influence between agents in the context of social media. We then define the following subtypes of this generic relationship:

- *prov-said:FollowRelationship* denotes that one agent was influenced by another agent, by establishing a unidirectional (follow) relationship. In the context of social media, that practically means being exposed to the messages emitted by the latter. For example, in Twitter this is the only way of connecting with users, while Facebook apart from the bidirectional *Friendship*, also gives the possibility of unidirectional connection by subscribing to the messages of users. Here, we assume that once an agent starts to follow another agent, he is exposed to his old messages and his future ones (if there are any). This is also the case in Twitter and Facebook. As a result, we do not model the influence that derives from exposure/subscription to messages explicitly, since it is implied by the *prov-said:FollowRelationship*.
  Furthermore, such a relationship entails a certain degree of uncertainty, since it can not be asserted whether an agent has seen the messages of another in reality.
- *prov-said:InteractionInfluence* denotes that an agent was influenced by another agent through interaction – i.e., by quoting, revising, replying to, or expressing emotions towards messages of the latter, or by mentioning the latter. The message types give us more information on which type of interaction occurred. As we explain in Sect. 5, this type of influence can possibly be discovered through the functionality of social media platforms or through message similarity.
- *prov-said:ExternalInfluence* denotes that an agent was influenced by a possibly unknown user and/or external event. Given that events that happen in reality are reflected in social media, we encounter messages that share certain similarity, but are not revised messages. An external event or meme drives the similarity in most of these cases. Given its prominence in social media, we decide to model external influence explicitly, which also allows us to differentiate it from message revision. An indicator that distinguishes these two cases might be the existence of social connection: in case that two messages share certain similarity and their authors are not connected, there is high probability of external influence. However, it is often the case that users

forward messages from the public timeline without the existence of a social relationship.

– *prov-said:SelfInfluence* denotes that an agent was influenced by himself. In social media we often encounter cases that a user is emitting similar messages in order to promote again specific content, or to edit messages.

We illustrate the influence types in Example 4.

Example 4: Influence Types

```
// User @Alice followed user @Carol, so a prov-said:FollowRelationship
// existed between them
prov:wasInfluencedBy(twitter:Alice, twitter:Carol,
   [prov:type='prov-said:FollowRelationship'])

// User @Bob revised a message from @Alice, so a
//prov-said:InteractionInfluence existed between them
prov:wasInfluencedBy(twitter:Bob, twitter:Alice,
   [prov:type='prov-said:InteractionInfluence'])

// User @Bob was influenced by an external event
// and the influencing agent is not specified
prov:wasInfluencedBy(twitter:Bob, -,
   [prov:type='prov-said:ExternalInfluence'])

// User @Bob was influenced by himself by propagating
// similar messages, as a result the agents of
// influencer and influencee are the same
prov:wasInfluencedBy(twitter:Bob, twitter:Bob,
   [prov:type='prov-said:SelfInfluence'])
```

By following these influence types, both the social graph and the interaction graph [34] can be reconstructed at a certain point in time by using provenance. The interaction graph aggregates interactions (e.g., emission of messages) among users as weighted edges.

4.2   Influence Activities

Additionally to the influence types expressed as relationships among users (subtypes of *prov:Influence*), we explicitly model the corresponding activities. This design decision offers greater expressiveness by providing more information about the time the influence relationships started and ended, what triggered them, etc. For these purposes, we introduce three subtypes of *prov:Activity*:

– *prov-said:InfluenceActivity* is subtype of prov:Activity. It denotes the activity of one agent influencing another with the following four subtypes:
– *prov-said:FollowActivity* denotes the activity of one agent to establish a unidirectional connection with another. Once such an activity starts, the first agent is exposed to the (future and past) message emissions of the latter.

This activity has a start time that denotes the time of establishing the connection and an optional end time in case the agent removes the connection with regard to the other agent.

– *prov-said:InteractionInfluenceActivity* denotes the activity of one agent influencing another, so that the latter reacts to the messages of the first. Note here that the *prov-said:InteractionInfluenceActivity* is instantaneous, and thus has the same start and end time (when the influenced message was emitted). In this way, we are able to model multiple interactions of agents by generating multiple instances of *prov-said:InteractionInfluenceActivity*. If we had considered the opposite case when *prov-said:InteractionInfluenceActivity* is asserted only once and has no end time, we would have come to contradiction with the principles of information diffusion, where the significance of past interactions fades quickly over time.

– prov-said:ExternalInfluenceActivity denotes the activity of an agent being influenced by an external, possibly unknown entity.

– prov-said:SelfInfluenceActivity denotes the activity of an agent influencing him or herself.

Example 5 illustrates the subtypes of *prov-said:InfluenceActivity*.

Example 5: Influence Activities

```
// A prov-said:FollowActivity was started at the moment user @Alice followed
// user @Carol. Since @Alice was still following @Carol at the time of assertion,
// there is no end time for the activity.
activity(alice-follows-carol, 2015-01-09T13:00:00, - ,
    [prov:type='prov-said:FollowActivity'])

// A prov-said:InteractionInfluenceActivity was started (and ended)
// at the moment user @Bob modified and re-emitted the message of @Alice.
activity(bob-influencedby-alice, 2015-01-09T13:05:00, 2015-01-09T13:05:00,
    [prov:type='prov-said:InteractionInfluenceActivity'])
wasStartedBy(bob-influencedby-alice, bob-status:23456, emit-23456,
    2015-01-09T13:05:00)
wasEndedBy(bob-influencedby-alice, bob-status:23456,
    emit-23456, 2015-01-09T13:05:00)
```

4.3   Influence Roles

Analysis of information diffusion and influence in social media makes use of specific roles for agents [2]. To model this, we need to specifically define values for the *prov:role* attribute in the context of *prov:Usage* and *prov:Association*. This way, we clarify the roles of agents involved in a *prov-said:InfluenceActivity*. We define the following role-values:

– *prov-said:Influencer* denotes the role of an agent that was used by an *prov-said:InfluenceActivity* that was associated with another agent. This means that the first agent influences the latter.

– *prov-said:Influencee* denotes the role of an agent that was associated with an *prov-said:InfluenceActivity*. This agent is being influenced by another agent used by the same *prov-said:InfluenceActivity*.

Following that, we define two subtypes of *prov-said:Influencee* and *prov-said:Influencer* respectively. Firstly, we model the follow relationship with the roles: *Follower* and *Followee* and secondly we model the activity of interaction by exchanging messages with the roles: *InteractionInfluencee* and *Interaction-Influencer*. Note that these roles are pairwise complementary by revealing the active behaviour of one agent in to establish social connections and to react to messages (follower, interactionInfluencee) and the passive behaviour of another (followee, interactionInfluencer) who exerts some influence on the first.

– *prov-said:Followee* denotes the role of an agent that was used by a *prov-said:FollowActivity* associated with another agent. This means that the latter followed the first.
– *prov-said:Follower* denotes the role of an agent that was associated with an *prov-said:FollowActivity*. This means than an agent establishes a unidirectional connection with another agent in social media.
– *prov-said:InteractionInfluencer* denotes the role of an agent that was used by an *provsaid:InteractionInfluenceActivity* associated with another agent. This means that the first agent is influencing the latter so that the latter reacts to the messages of the first.
– *prov-said:InteractionInfluencee* denotes the role of an agent that was associated with an *prov-said:InteractionInfluenceActivity*. This means that the agent is being influenced by another agent by reacting to the messages of the latter.

We demonstrate these roles in Example 6.

Example 6: Influence Roles

```
used(alice-follows-carol, twitter:Carol, [prov:role='prov-said:Followee'])
wasAssociatedWith(alice-follows-carol, twitter:Alice,
   [prov:role='prov-said:Follower'])

used(bob-influencedby-alice, twitter:Alice,
   [prov:role='prov-said:InteractionInfluencer'])
wasAssociatedWith(bob-influencedby-alice,  twitter:Bob,
   [prov:role='prov-said:InteractionInfluencee'])
```

Note here that we do not define a specific *prov:role* for the influence types *prov-said:ExternalInfluence* and *prov-said:SelfInfluence*. We argue that the more generic *prov-said:Influencer* and *prov-said:Influencee* suffice to model the roles in these cases. The type of Influence is demonstrated by the type of relationship and activity.

At this point we express the following constraints:

- We ensure that an *prov-said:InfluenceRelationship* always implies a *prov-said:InfluenceActivity*, *prov:Usage* and *prov:Association*. According to the type of *prov-said:InfluenceActivity*, specific *prov:roles* are being used.
- A *prov-said:InteractionInfluenceActivity* starts (and ends, since it is defined to be instantaneous) with the emission of *prov-said:CopiedMessage*, *prov-said:RevisedMessage*, *prov-said:ReplyMessage*, *prov-said:EmotionMessage* or *prov-said:MentionMessage*
- A *prov-said:SelfInfluence* implies that the *prov:Agents* that influence each other are the same.

With these concepts, we have covered the model of influence with its possible expressions in activities, relationships and roles.

5   Methods for Provenance Generation

In this section, we present methods to generate actual provenance data which are mapped to our model. While the *PROV-SAID* model remains general enough to express multiple types of user interactions on different platforms, we present the reconstruction methods on Twitter. Twitter is one of the most studied social media, offering rich possibilities for user interactions, both explicit and implicit. Similar methods can be applied to other social media platforms, Q&A forums or news media, which we also will evaluate in Sect. 6. We extend and combine our works for provenance reconstruction from [11,30,32]. In particular, we develop and present methods to reconstruct replies and quotes in Twitter that complements our previous work for retweet [30] and implicit provenance reconstruction [11,32].

As discussed, we discern two types of interactions in social media. The first one is *explicit*, where users are leveraging social media embedded functions and create provenance information as metadata. The second one is *implicit*, where users are influenced by others, but do not expose such provenance explicitly to the platform. In this case, they might be using their own conventions to express influence or credit implicitly, e.g., by mentioning the influencer in the message content. Yet, such provenance information is not captured by the metadata of social media providers. For these reasons, any reconstructed provenance carries uncertainty.

5.1   Explicit Interactions

For *explicit interactions*, the functions that each social platform provides are used, such as retweets and quotes in Twitter, shares in Facebook, replies, etc. That way, users are showing credit to preceding message(s) which can be captured as provenance information. However, social media platforms fail to fully expose such provenance: for retweets, the root of the retweet cascade is provided, but not the intermediate steps; for replies and quotes, the previous step is exposed, but not the root[4]. The combination of such features further complicates matters: for example when a user quotes a retweet, the retweet root gets lost. Here, we leverage the provenance information that is provided from Twitter to reconstruct provenance and find the intermediate steps and/or the root of diffusion. We use two methods which can be applied to other social platforms:

- *Iterative Traversal of Attributions* in case the previous step is provided: we unfold the diffusion paths with the aim of identifying the root and also the other derived messages, if possible.
- *Social connections* in case the previous step(s) are not being exposed: we reconstruct the diffusion paths under the hypothesis that information flow through social connections [30].

These will be explain in more detail in the next Sections.

Reconstructing Replies and Quotes Through Iterative Traversal. For reconstructing conversations (reply cascades) in Twitter, we use iteration to identify provenance. As discussed in the case of replies, the previous step is given but not the root of the conversation. As a result, reply cascades are trees, since Twitter API provides only the previous influencer. This way, we iteratively reconstruct diffusion paths until we reach the conversation root (*backward search*). In order to collect additional replies (further branches of the reply tree that are not included to the crawled dataset), we refer to the Twitter Timeline. The Twitter API allows us to get further replies from these users who participated in the conversation (*forward search*), by collecting their timelines. The search can retrieve approximately 3000 messages for each timeline. As a result, we can get a good coverage for users that are not very active.

Forward search is very useful for most of our use cases. We are interested in reconstructing the provenance of data that refer to particular events (Olympics, Elections, Conferences, etc) by collecting datasets containing specific keywords. However, replies might not always contain these keywords and as a result they are not captured from our crawler. By the forward search, we are able to collect additional replies and complement the conversation trees.

Quotes are being reconstructed in the equivalent way as replies since their attribution is represented similarly. Technically there is a small difference in retrieving additional quotes (forward search): Additional quotes cannot be retrieved using the timeline API, so that we need to use the Search API with the message identifier of the quoted message as a key. This API allows retrieval for only around one week into the past, which means that such provenance needs to be reconstructed in a timely way.

The empirical observation of the combination of replies and quotes has brought the insight that a tweet may contain multiple attributions, namely a reply and a quote. This turns the attribution tree with a single root node into an attribution DAG (directed acyclic graphs) containing multiple "roots".

---

[4] Note here that a quote in Twitter is a *prov-said:RevisedMessage* and a retweet is a *prov-said:CopiedMessage*.

Reconstructing Retweets Through Social Connections. For reconstructing retweet cascades we use our method from [31] which leverages the social graph in order to reconstruct the intermediate diffusion paths and find possible influencers. Recall that in the case of retweets, the Twitter API provides the root and not the intermediate steps. We rely on the assumption that information flows over the social graph and users are influenced by their connections in order to propagate a piece of information. Note that there might exist multiple influencers, in cases where more than one connection is activated. In this case retweet cascades are DAGs with a single root node, yet multiple paths from a node to this root.

5.2   Implicit Interactions

For *implicit interactions*, users are not exposing any type of provenance through social media functions, and as a result provenance is lost (e.g., when a user copies a message instead of retweeting it). We leverage two approaches to reconstruct such provenance: the first based on content similarity in Sect. 5.3, the second as an extension that utilizes additional indications and user conventions in Sect. 5.3.

5.3   Similarity Algorithm

To reconstruct this kind of information diffusion, we adapt our approach from [11] to reconstruct fine-grained provenance. The core assumption of this approach is: *"if two messages are highly similar, there is a high probability that they share some provenance"*. The adapted approach consists of the following steps:

1. Remove all explicitly copied messages keeping only the original messages;
2. Index the text content of this reduced dataset after tokenization (including stopwords) and stemming using a feature model and semantic similarity function (e.g., TF-IDF and the cosine similarity between the term vectors of the messages), and compute the full similarity matrix of all messages;
3. Apply a similarity-based clustering algorithm such as SimClus [1] to divide the dataset into (possibly overlapping) clusters of messages that all have a similarity to each other higher than a predetermined threshold;
4. For each cluster:
    - identify the oldest message as the root message of that cluster;
    - connect all other messages to the root message by a
      prov-said:IndirectDerivation
        • if the message is identical to the root message, using a
          prov:wasQuotedFrom relationship;
        • if the message is not identical to the root message, using a
          prov:wasRevisionOf relationship.
    - determine possible intermediate steps by matching each message to its most similar (highest cosine similarity) by *prov:Derivation* respecting the temporal order. Again:

- if the message is identical to its most similar predecessor, using a *prov:wasQuotedFrom* relationship;
- if the message is not identical to its most similar predecessor, using a *prov:wasRevisionOf* relationship.

Using these steps, we enrich the dataset and expose previously hidden knowledge about information diffusion. The main idea is that all messages are indirectly derived through *any step* from the oldest message in the cluster and through one *single step* from their most similar predecessor message, the latter providing possible intermediate steps.

We have scaled and optimized this algorithm in order to compute similarity over infinite streams [33] and we refer the reader here [33] for further information. The technical details of this algorithm are therefore out of the scope of this paper.

Additional Indicators: User Conventions. In addition the similarity-based provenance identified by the aforementioned algorithm, extensive manual inspection and evaluation led us indicators that reveal additional provenance or decrease the uncertainty of the reconstructed provenance. These indicators capture user interaction conventions that commonly occur in Twitter and support the following types of influence: (1) user influence: (1.a) with explicit credit, (1.b) without credit, (2) external influence, (3) self-influence: (3.a) delete and rewrite, (3.b) promotion.

(1) As a first case, we observed that a user is influenced by another user by propagating similar messages. Here, two cases can occur. (1.a) Sometimes, users prefer to give explicit credit to the initial contributor by mentioning the username within the message text (with "@" or "via"). This behaviour was adopted by users before the retweet feature was released. (1.b) In case there is no explicit credit within the message text and there is still high similarity between two messages, we check if the users are neighbours in the social graph. Since users are exposed to the messages by their connections, there is a high chance that they are influenced by them.

(2) If no social graph connection exists between two users that are emitting highly similar messages, we observed that there is either an external event that drives the similarity (e.g., a football match) or influence from public trends.

(3) As a last case, we observed that many highly similar messages share the same author. For Twitter, we discerned two categories: (3.a) highly similar messages from the same author with the earlier being deleted, as the result of users who delete and rewrite their own messages, since there is no edit feature provided by Twitter; (3.b) promotion of existing information by the same user, e.g., for advertisement. Figure 3 depicts the workflow we use to discern these cases. Since these cases are covered by the extended *PROV-SAID* model, more specific observations tied to Twitter are not modeled explicitly.

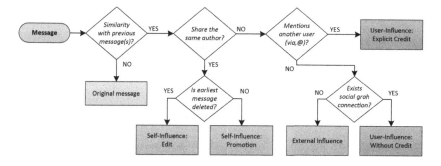

Fig. 3. Flow diagram for implicit interactions

Note here that in any case implicit diffusion entails higher uncertainty provenance compared to explicit diffusion. We chose to express uncertainty as a function of the similarity shared between two messages, the existence of shared social connections, the temporal and spatial distance, etc. However, identifying the uncertainty of provenance asserted by the *PROV-SAID* model is out of scope for this paper, and part of our future work.

6   Evaluation

For evaluating our model and assumptions, we computed both implicit and explicit provenance on several datasets from different source types and varying in size. We are presented an in-depth analysis of a Twitter dataset and higher-level analysis of a newsfeed dataset here. A more extended evaluation (also covering performance aspects) can be found in [33].

6.1   Twitter Dataset

To analyze interactions in Twitter, we used a small and controlled dataset so that we could perform an in-depth study. This data set was recorded during the World Wide Web Conference in 2015 (WWW2015). It consists of 7363 messages in total, of which 3819 are retweets, 141 are replies and 22 are quotes.

Explicit Interactions. For *explicit* interactions we reconstructed retweet, reply and quote cascades according to the methods in Sect. 5.1. The provenance edges created by explicit interactions carry high certainty, since provenance information is provided by Twitter as metadata in the collected messages.

We identified 1162 retweet cascades with size distribution presented in Fig. 4: Very few have a large size (the largest consists of 109 retweets and the high majority contains 1–2 retweets). Retweets represent *prov-said:CopiedMessages* in *PROV-SAID*. Replies and quotes constitute a very small fraction and they are often observed interleaved in the same cascade. We identified 114 reply cascades showing a skewed distribution with the largest reply cascade having 53

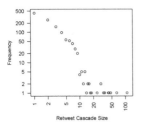
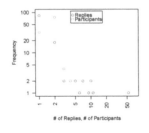
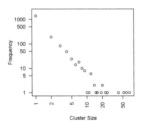

Fig. 4. Retweet cascade Distribution

Fig. 5. Reply cascade Distribution

Fig. 6. Cluster distribution

replies and the majority obtaining 1–2 replies (Fig. 5). We also plotted the distribution of the number of distinct users participating in the conversations. We observe that the majority consists of two participants, which means that users are highly conversational in this dataset. Quotes represent only a tiny fraction of the dataset: the largest cascade contains 8 quotes and the remainder 1–2 quotes. At that time, quotes were a new feature and relatively unknown to most Twitter users. Replies represent *prov-said:ReplyMessage*s while quotes represent *prov-said:RevisedMessage* (Fig. 6).

Implicit Interactions. After computing the provenance of explicit interactions, we identify *implicit* provenance where no additional information is provided from Twitter. For computing such provenance we use our algorithm from [11] presented in Sect. 5.2. The dataset was cleaned from retweets, since their content is identical, while keeping the roots of the retweet cascades as *original messages*. We included replies and quotes because they consist of possibly new content and might carry additional implicit provenance. We end up with 3262 distinct messages.

Next we apply the similarity algorithm from Sect. 5.2 which takes as an input the lower bound of similarity. We compute the RMMSTD metric [20] that assesses the quality of clustering with different similarity thresholds. We prove that for similar datasets, appropriate thresholds range from 0.4–0.9. However, thresholds of 0.8–0.9 result in a large number of singletons which is not the desired goal. Additionally, the sparser the dataset, the lower we select the similarity threshold so that we end up with sufficient number of provenance edges. By evaluating manually the provenance edges generated from the similarity algorithm, we obtain high values of precision. We cannot compute recall, since we have to evaluate all combinations of possible connections, also outside of our crawled dataset.

We apply the similarity algorithm with lower bound of similarity 0.4, as it provided the densest results set while still providing adequate cluster quality metrics. The algorithm identified 415 overlapping clusters that also correspond to the different topics discussed during WWW2015. 1441 messages were singletons since they did not demonstrate similarity higher then the threshold with any other message. Within these clusters, 841 indirect derivations (*prov-said:IndirectDerivation*s) and 864 direct derivations (*prov:Derivation*s) were identified. The number of indirect and direct derivations are by definition close since we are computing two connection for every user as explained in Sect. 5.2 (the oldest in the cluster and the most similar). These numbers indicate that we could identify implicit provenance for 24% of the messages in our dataset (excluding retweets). Note that this number is likely overestimated since there is *external influence* in this particular dataset. For example, participants of the conference might emit messages referring to conference presentations which share limited provenance with other messages.

On top of these provenance edges we leverage the indicators from [32] presented in Sect. 5.2 where we strive to increase the certainty of these provenance edges. We observed that 315 messages share the same author, out of which 127 messages substitute the edit function of Twitter (users deleting and rewriting messages) and 188 are promoting the same content (*prov-said:SelfInfluence*). We observe that almost 10% of interactions can be explained by just looking at the individual users' behavior.

For explicit mentions, we identified 105 messages that propagate the content of a message by mentioning the original author. This (high) number may be specific to this (often conversational) dataset, as most of the authors are computer scientists possibly adhering to the old Twitter conventions of propagating content. We also investigate the social graph connection between these users. The majority (45%) of such connections are bidirectional (both users follo each other). However, there is a surprising 17% of messages where a follower is credited by a followee through a mention. Lastly, 38% of users share no relationship with the user whom they mention. The last fact contradicts the assumptions of [5] that users are influenced by their connections. Complementary, 42 users influenced others by exposing relevant content through mention. This type of influence (*influence by mentioning*) was identified in [33].

These provenance edges with additional indicators (in total 462 or 14%) carry high certainty provenance. The remaining 10% (out of 24% from the result of clustering) represent either external influence or influence without any obvious indicators: these two types of influence is hard to differentiate without any event identification algorithm.

All the aforementioned provenance computations are expressed in RDF through [12] and comply fully with the *PROV-SAID* model. The results are summarized in Table 1.

Table 1. Explicit and implicit provenance for WWW2015 dataset

Explicit #	Implicit #
Retweet: 3819	Indirect derivations: 864
Replies: 141	Self-influence: 315
Quotes: 22	Explicit credit: 42
	Influence by mention: 105

6.2 Newsfeed Dataset

Our second dataset contains messages from a news aggregator[5]. It consists of 65K news articles in English covering the first 12 h on 02/01/17. By applying the similarity algorithm and setting the similarity threshold at 0.5, we identified 3265 clusters, which nicely cover the different topics of this dataset. The topics include the North Carolina's transgender bathroom law, the expulsion of 35 suspected Russian spies, an armed attack in a Bahrain prison, sport's events, etc. We also identified 2020 one-step derivations which shows the provenance of articles. Updates could be matched to the previously published articles and republishing of the same article could be linked to the original sources. Here, there is no underlying social graph and no user interactions as in the case of Twitter. As a result, we do not compute the additional indicators on top of implicit influence, which are targeting mainly social media use cases.

The main bottleneck of the similarity algorithm is the similarity computation of documents. However, it has acceptable performance for the news-articles dataset: the documents are much larger than tweets, but the frequency is much lower than the twitter stream. In general, we observe two spectrum were most of real life datasets lie: short, fast-paced documents (social media), or long, low frequency documents (news articles). We have proved that the algorithm works efficiently and produces reasonable results for both cases.

7 Related Work

While *Information Diffusion* in social media has received a lot of attention, in particular its modeling [15], there is limited work on the reverse procedure, i.e. *information provenance*, which is the focus of this paper. We divide the state-of-the-art in this area in the following main two categories: (i) Models for information diffusion and provenance in social media and use cases, (ii) Computational methods

Information Diffusion in social media and networks has been a well researched topic. A review of relevant models can be found in [15]. Until now, the focus has been on the design of models with specific goals [4], e.g. assessing the probability that certain users are being reached. Such research is mostly driven by data mining techniques, (algorithms, frameworks and systems) to analyze specific datasets. This sort of analysis is useful in use cases such as marketing, for instance solving the problem of maximizing the spread of information by targeting a specific users (i.e., the *influence maximization* problem [19]).

PROV-DM has been applied to various use cases. For example, authors in [25] used this model to express history of revisions in Wikipedia. The work in [10] provides a mapping of *PROV-DM* to the version control system of Git. The produced provenance data can be consumed by other tools using *PROV*. In a different domain, authors model the history of clinical guidelines with *PROV* [23].

---

[5] http://newsfeed.ijs.si.

Such work facilitates the understanding of provided recommendations by practitioners. More use cases are listed in the *PROV* implementation report [18], and the list continues to grow.

PROV was deliberately kept as generic and extensible as possible, to allow for many application fields. As a result, many extensions have been proposed to express the semantics of particular use cases. For example, in the context of neuroimaging, the *PROV-DM* was extended in order to capture provenance between the stages that neuroimaging data undergoes: data acquisition, pre-processing and statistical analysis [24]. This way, relevant parameters used in each stage can be traced back which facilitates reproducibility and metadata analysis. A general extension to *PROV-DM* was proposed by [8] in order to capture the concept of uncertainty in two ways: uncertainty in provenance statements and uncertainty about the content of an entity whose provenance is assessed. This last extension is very useful when algorithms with a certain degree of uncertainty are used to assert the provenance. For example, if an information diffusion detector was used to assert the *PROV* graphs mentioned throughout this paper, UP [8] would be a way to annotate the provenance with the detector's confidence.

For computing provenance of in social media data, we identified the following methods: (i) provenance through content similarity; (ii) provenance through social graph connections; (iii) provenance through user profile metadata.

(i) The work of [22] focuses on tracing news and quotes (referred to as *memes*) on the Web over time by revealing temporal patterns, mutations (alterations) that online phrases undergo and properties of the news' life cycle. A subsequent work using the same datasets and methods [28] shifts the focus on fine-grained content alterations. The research in [5], computes influence in information diffusion (especially replies) that official mechanisms cannot capture in Twitter. While such work is close to our provenance computations through similarity, the assumption in [5] is different from ours: users are being influenced by the last 100 messages from their friends on their timeline. However, according to our empirical evaluations, we have observed influence among users (e.g., mentions and manually copied messages) without any obvious social graph connection.

In [9], the provenance of news articles is reconstructed automatically using semantic similarity. In this paper, we adapt this approach for social media and the *PROV-SAID* model.

(ii) Traditional information diffusion research includes tracing a piece of information back to its sources through social connections, revealing the concepts of influence and trust among the users involved. The work of [13] recovers information recipient sub-graphs given a small fraction of known recipients. In [16] unknown recipients are identified under the assumptions of degree and closeness propensity: nodes with a higher degree and closer to the sources are more likely to propagate information. [4] provides a provenance reconstruction method through social connections based on well established information diffusion models. Finally, in [30], we automatically reconstruct *information cascades* that show the paths information has flown, given a piece of information that propagates over a social graph. Information cascades are graphs that model how information is being diffused from user to user; in other words, our approach in [30] reconstructs the paths of users who propagate information back to the sources by finding intermediate influencers.

(iii) Lastly, provenance can be derived through user profile metadata, attributing relevance and trust to the information emitted according to the characteristics of the contributor. The work of [17] implements a tool for collecting such user information from different media, while not providing any information on the provenance paths and sources.

For computing provenance, we combine concepts from (i), (ii) and our empitical observations in order to reveal and express provenance paths, by extending and adapting the solutions proposed in [9, 11, 30, 32]. Finally, the results are modeled and combined in an interoperable way using the *PROV-SAID* model.

8 Conclusion and Future Work

To sum up, *PROV-SAID* enables systems that analyze social media to incorporate provenance data in their information diffusion analysis. This will benefit the massive human-centric efforts for judging relevance and trustworthiness of information by unraveling its sources and intermediate steps. In this paper, we extend our *PROV-SAID* model to account for multiple types of interactions. We present methods to compute such provenance in Twitter that can be applied in any social media or Q&A platform. We empirically identified indicators that decrease the uncertainty in the reconstructed provenance.

For future work, we plan to investigate computational methods for quantifying uncertainty. Here, we present some relative ranking for uncertainty among the provenance edges; however quantifying uncertainty is a challenging problem where contextual information must be taken into account, e.g. spatiotemporal information.

References

1. Al Hasan, M., Salem, S., Zaki, M.J.: SimClus: an effective algorithm for clustering with a lower bound on similarity. Knowl. Inf. Syst. **28**(3), 665–685 (2011)
2. Bakshy, E., Hofman, J.M., Mason, W.A., Watts, D.J.: Everyone's an influencer: quantifying influence on Twitter. In: WSDM, pp. 65–74 (2011)
3. Baños, R.A., Borge-Holthoefer, J., Moreno, Y.: The role of hidden influentials in the diffusion of online information cascades. EPJ Data Sci. **2**(1), 1–16 (2013)
4. Barbier, G., Feng, Z., Gundecha, P., Liu, H.: Provenance data in social media. Synth. Lect. Data Min. Knowl. Discov. **4**(1), 1–84 (2013)
5. Barbosa, S., Cesar-Jr, R.M., Cosley, D.: Using text similarity to detect social interactions not captured by formal reply mechanisms. In: 2015 IEEE 11th International Conference on e-Science (e-Science), pp. 36–46. IEEE (2015)

6. Cheney, J., Chiticariu, L., Tan, W.-C.: Provenance in Databases: Why, How, and Where, vol. 4. Now Publishers, Inc., Hanover (2009)

7. Cheney, J.: W3C Provenance Working Group: Semantics of the PROV Data Model. W3C Note, 30 April 2013

8. De Nies, T., Coppens, S., Mannens, E. Van de Walle, R.: Modeling uncertain provenance and provenance of uncertainty in W3C PROV. In: WWW (Companion Volume), pp. 167–168 (2013)

9. De Nies, T., Coppens, S., Van Deursen, D., Mannens, E., Van de Walle, R.: Automatic discovery of high-level provenance using semantic similarity. In: Groth, P., Frew, J. (eds.) IPAW 2012. LNCS, vol. 7525, pp. 97–110. Springer, Heidelberg (2012). https://doi.org/10.1007/978-3-642-34222-6_8

10. De Nies, T., et al.: Git2PROV: exposing version control system content as W3C PROV. In: ISWC (Posters & Demos), pp. 125–128 (2013)

11. De Nies, T., et al.: Towards multi-level provenance reconstruction of information diffusion on social media. In: Proceedings of the 24th ACM International on Conference on Information and Knowledge Management, pp. 1823–1826 (2015)

12. Dimou, A., Vander Sande, M., Colpaert, P., Verborgh, R., Mannens, E., Van de Walle, R.: RML: a generic language for integrated RDF mappings of heterogeneous data. In: Proceedings of the 7th Workshop on Linked Data on the Web (LDOW 2014), Seoul, Korea (2014)

13. Feng, Z., Gundecha, P., Liu, H.: Recovering information recipients in social media via provenance. In: ASONAM, pp. 706–711 (2013)

14. Glavic, B., Sheykh Esmaili, K., Fischer, P.M., Tatbul, N.: Ariadne: managing fine-grained provenance on data streams. In: Distributed Event-Based Systems, pp. 39–50 (2013)

15. Guille, A., Hacid, H., Favre, C., Zighed, D.A.: Information diffusion in online social networks: a survey. ACM SIGMOD Rec. **42**(2), 17–28 (2013)

16. Gundecha, P., Feng, Z., Liu, H.: Seeking provenance of information using social media. In: CIKM (2013)

17. Gundecha, P., Ranganath, S., Feng, Z., Liu, H.: A tool for collecting provenance data in social media. In: KDD, pp. 1462–1465 (2013)

18. Huynh, T., Groth, P., Zednik, S. (eds.) and W3C Provenance Working Group: PROV Implementation Report. W3C Working Group Note, 30 April 2013

19. Kempe, D., Kleinberg, J., Tardos, É.: Maximizing the spread of influence through a social network. In: SIGKDD pp. 137–146 (2003)

20. Kovács, F., Legány, C., Babos, A.: Cluster validity measurement techniques. In: 6th International Symposium of Hungarian Researchers on Computational Intelligence. Citeseer (2005)

21. Kwon, S., Cha, M., Jung, K., Chen, W., Wang, Y.: Prominent features of rumor propagation in online social media. In: 2013 IEEE 13th International Conference on Data Mining (ICDM), pp. 1103–1108. IEEE (2013)

22. Leskovec, J., Backstrom, L., Kleinberg, J.: Meme-tracking and the dynamics of the news cycle. In: KDD, pp. 497–506 (2009)

23. Magliacane, S., Groth, P.T., et al.: Towards reconstructing the provenance of clinical guidelines. In: SWAT4LS (2012)

24. Maumet, C., Flandin, G., Nichols, B., Steffener, J., Helmer, K., et al.: Extending NI-DM to share the results and provenance of a neuroimaging study: implementation within SPM and FSL. Front. Neuroinform. (2014)

25. Missier, P., Chen, Z.: Extracting PROV provenance traces from Wikipedia history pages. In: EDBT/ICDT (Workshops), pp. 327–330 (2013)

26. Moreau, L., Missier, P. (eds.) and W3C Provenance Working Group: PROV-DM: The PROV Data Model. W3C (2013)
27. Myers, S.A., Zhu, C., Leskovec, J.: Information diffusion and external influence in networks. In: SIGKDD, pp. 33–41 (2012)
28. Simmons, M.P., Adamic, L.A., Adar, E.: Memes online: extracted, subtracted, injected, and recollected. ICWSM **11**, 17–21 (2011)
29. Taxidou, I., De Nies, T., Verborgh, R., Fischer, P.M., Mannens, E., Van de Walle, R.: Modeling information diffusion in social media as provenance with W3C PROV. In: Proceedings of the 24th International Conference on World Wide Web, pp. 819–824 (2015)
30. Taxidou, I., Fischer, P.M.: Online analysis of information diffusion in Twitter. In: Proceedings of the 23rd International Conference on WWW Companion, pp. 1313–1318 (2014)
31. Taxidou, I., Fischer, P.M.: Online analysis of information diffusion in Twitter. In: WWW (Companion Volume), pp. 1313–1318 (2014)
32. Taxidou, I., Fischer, P.M., De Nies, T., Mannens, E., Van de Walle, R.: Information diffusion and provenance of interactions in Twitter: is it only about retweets? In: Proceedings of the 25th International Conference Companion on World Wide Web, pp. 113–114 (2016)
33. Taxidou, I., Lieber, S., Fischer, P.M., De Nies, T., Verborgh, R.: Web-scale provenance reconstruction of implicit information diffusion on social media. Under Review (2017)
34. Wilson, C., Sala, A., Puttaswamy, K.P., Zhao, B.Y.: Beyond social graphs: user interactions in online social networks and their implications. ACM Trans. Web **6**(4), 17 (2012)

Author Index

Printed in the United States
By Bookmasters